COGNITION AND COMPUTERS:
Studies in learning

ELLIS HORWOOD SERIES IN COGNITIVE SCIENCE

Series Editor: MASOUD YAZDANI, Department of Computer Science, University of Exeter

R.W. Lawler	**Computer Experience and Cognitive Development A Child's Learning in a Computer Culture**
R.W. Lawler, B. du Boulay, M. Hughes & H. Macleod	**Cognition and Computers: Studies in Learning**
A. Narayanan	**On Being A Machine**
N.E. Sharkey	**Advances in Cognitive Science 1**
M. Sharples	**Cognition, Computers and Creative Writing**
Sloman, A.	**On Cognition and Computation**

COGNITION AND COMPUTERS:

Studies in learning

ROBERT W. LAWLER
Fundamental Research Lab, GTE Laboratories, Waltham, MA, USA
BENEDICT DU BOULAY
Cognitive Studies Programme, University of Sussex, UK
MARTIN HUGHES
School of Education, University of Exeter, UK
and
HAMISH MACLEOD
Department of Psychology, University of Edinburgh, UK

ELLIS HORWOOD LIMITED
Publishers · Chichester

Halsted Press: a division of
JOHN WILEY & SONS
New York · Chichester · Brisbane · Toronto

First published in 1986 by
ELLIS HORWOOD LIMITED
Market Cross House, Cooper Street, Chichester, West Sussex, PO19 1EB, England

The publisher's colophon is reproduced from James Gillison's drawing of the ancient Market Cross, Chichester.

Distributors:
Australia and New Zealand:
JACARANDA WILEY LIMITED
GPO Box 859, Brisbane, Queensland 4001, Australia
Canada:
JOHN WILEY & SONS CANADA LIMITED
22 Worcester Road, Rexdale, Ontario, Canada
Europe and Africa:
JOHN WILEY & SONS LIMITED
Baffins Lane, Chichester, West Sussex, England
North and South America and the rest of the world:
Halsted Press: a division of
JOHN WILEY & SONS
605 Third Avenue, New York, NY 10158, USA

British Library Cataloguing in Publication Data
Cognition and computers: studies in learnilng. —
(Ellis Horwood series in cognitive science)
1. Cognition in children
2. Human information processing in children
I. Lawler, R. W.
155.4'13 BF723.C5

Library of Congress Card No. 86–2928

ISBN 0–7458–0006–8 (Ellis Horwood Limited — Library Edn.)
ISBN 0–7458–0049–1 (Ellis Horwood Limited — Student Edn.)
ISBN 0–470–20324–2 (Halsted Press — Library Edn.)
ISBN 0–470–20328–5 (Halsted Press — Student Edn.)

Phototypeset in Times by Ellis Horwood Limited
Printed in Great Britain by Unwin Bros. of Woking

Contents

Contents

Foreword

Can computers help cognitive development? Seymour Papert, in his seminal book *Mindstorms*, presented a strong case for believing that the answer must be positive. But how can we know for sure? Papert simultaneously argued against attempting to use traditional evaluation methods when dealing with a fundamental phenomenon which could have far-reaching behavioural effects. His argument is summarized in an example of the role played by differential gears in his own intellectual development:

> First, I remember that no one told me to learn about differential gears. Second, I remember that there was *feeling, love,* as well as understanding in my relationship with gears. Third, I remember that my first encounter with them was in my second year. If any 'scientific' educational psychologist had tried to 'measure' the effects of this encounter, he would probably have failed. It had profound consequences but, I conjecture, only very many years later. A 'pre- and post-' test at age two would have missed them.

What links the research reported in the three parts of this book is the desire by the authors to investigate the claims made for the effects on an individual's cognitive development of computer education in its most enlighted Logo-based form.

Lawler, long a colleague of Papert, complements the case study of his daughter's development with a four chapter observation of his son's learning about numbers.

Du Boulay presents three case studies of student teachers who were observed in detail as they learned Logo and tried to apply it to various areas of mathematics in which they were weak.

Hughes and MacLeod evaluate a simplified version of Logo for use with very young children. They give extended examples from particular sessions working with the children.

I believe this book makes a major contribution to clarifying the role that computers can play in an individual's cognitive development. I also hope that the publication of the studies presented in this book leads to further work being undertaken to ascertain in detail the effect of using computational artifacts in education.

Masoud Yazdani

Exeter
January, 1986

Part 1

Natural learning: people, computers, and every-day number knowledge

Robert W. Lawler

In a time when Freud's suspicion of human motives has become the background against which all actions are seen, when we recognize, as Auden has it, that 'the desires of the heart are as crooked as corkscrews', any approach to knowledge which smacks of introspection or engaged observation is suspect as well. Yet Langer (1967) argues convincingly that only through exploring the engagement of observer and psychological subject will we ever come to confront those phenomena which are central to an appreciation of learning in the language capable mind.

The chapters in this part of the book focus on understanding natural learning though applications of the case method. the four studies all focus on the number-related learning of my son, Robby, during his sixth and seventh years. The other thread that binds them together is the issue of rendering more accessible and dependable those kinds of insights that can only come about through the articulate testimony of the psychological subject. The collection of ideas and observations move, in their attempt to deal with the issue of learning, from introspection to a balanced mixture of mechanically recorded experiment and naturalistic observation. The inspiration for beginning this recording was the observation by a fellow researcher† that a major technique being explored by Genevan psychologists involved the following steps:

— a video tape is recorded of a person solving a problem;
— the video tape is replayed, and the problem solver is asked to explain, in a second recorded session, what he was thinking when he did specific steps of the problem;
— the researchers then examine the most interesting portions of those tapes in exquisite detail.

†Laurence Miller, who had studied with G. Cellerier at Piaget's Center for Genetic Epistemology in Geneva.

At the laboratory where the following studies were undertaken,† I began to ask how we could apply and modify such techniques for use in the study of natural learning.

INTROSPECTION AND CASE STUDY METHODOLOGY

Generally, people's understandings of various problems depend on their own experience more than on general principles, but experience is characteristically particular as to content and episodic in structure. Introspection is vulnerable to criticism as a review of atypical experience where the details and significance of experiences are modified to conform to the individual's preconceptions. Capturing a record of one's own behavior for later analysis might permit the judicious mixture of reproducible records of experience with a richer appreciation of the possible meanings of the recorded behavior. Such an attempt at constrained, analytic introspection begins this collection in Chapter 1, 'Emerging form in turtle geometry'.

Obviously enough, introspection alone is inadequate for studies of instruction. The extension of constrained, analytic reflection to situations where ideas are transferred from one mind to another is explored in Chapter 2, 'Three encounters with number'. As an engaged experimenter, I brought to this study the possibility of understanding the behavior from the inside; further, as the father of the subject, I have a long familiarity with him which helps significantly in interpreting what he meant by what he said and did.

Given that our focus is on natural learning, in the sense of the collection of processes which create commonsense knowledge, it proved essential to move beyond the limits of experimenter-directed activities to explore how the initiative of the child and the social context reveal the character of natural learning in its interactions with people and problems. Chapter 3, 'Sketches of natural learning', is a selection of observations drawn from a more extensive but sparse collection of experimental protocols on various themes taken up with my son during his seventh year. Finally, Chapter 4, 'Extending a power idea', explores how theory construction derives from pre-existing heuristics for exploring and creating interesting new problems.

As a study of one child's learning in a computer-rich milieu, this collection is a companion piece to *Computer Experience and Cognitive Development* (Ellis Horwood, 1985), the more intensive study and analysis of my daughter's learning in her sixth year. Both studies took place in the same laboratory, at the same time. They both are, in fact, different aspects of the same effort to explore the processes underlying the development of learned, everyday knowledge.

ACKNOWLEDGEMENT

My thanks go to Masoud Yazdani for proposing that I assemble the early work about my son at MIT to complement the study of my daughter.

†The Children's Learning Lab of the MIT Logo Group. The Logo project was then a part of the MIT Artificial Intelligence Laboratory.

1

Emerging forms in turtle geometry

This chapter is an introspective study, focussed on the development of a simple but non-trivial idea, proceeding through an analysis of, and reflection upon, a trace of my own thought. It is a noting of the appearance of ideas, their sources, and interconnections. This retrospective approach has been possible because my habit of long standing has been to keep an intellectual journal.

The generative theme of this case study has been the idea of emergence. In a lecture focussed on the history of Artificial Intelligence, Marvin Minsky (1969) asked whether mental coherence in the human mind is a centrally generated phenomenon or an incidental artifact, emergent from the interaction of parts with clustering procedures. He remarked further that a rich theory of the kinds of emergence that can occur in the human mind is needed and that it is more likely to come from computer-related sciences than from mathematics; for the former confront the issue of complex interaction immediately, while the latter has failed to develop any profound method for dealing with the relation of wholes and parts At that time, I was sure I didn't know what Marvin meant; now I am less certain.

From my journal's notes of the ensuing month touching on the issue of emergence most entries related to language and in two aspects which are adequately characterized by the terms ontogenetic and phylogenetic. The latter motif springs from another observation of Minsky's, that 'evolutionary genesis' must be addressed by any theory which purports to explain the human mind. This observation led to my re-reading various works of Susanne Langer, which will be noted in more detail later. The former aspect, the ontogenesis of linguistic forms, takes its keynote from the two observations that what is needed is an explanation of structure as an emergent in linguistic development and that mathematics has failed to deal adequately with the problems of parts and wholes and their interactions. I note these observations not to raise arguments but to reflect my building sense of the importance of emergent processes and phenomena and consequently, a developing motivation to confront and explore such phenomena wherever they might appear accessible.

A CONCRETE EMBODIMENT OF THE PROBLEM

More than a month after the planting of this seed in my mind, the process by which it became embodied began with a computer session at the MIT Logo project. A colleague in a teacher's group asked me to discuss her work on the computer; she wanted someone to talk to so that she could could externalize her ideas, tape record them, and subsequently analyze their development. At the end of that session, we spent some time developing a procedure to draw some of the designs shown in Fig. 1.1.† At that point, I saw the

Fig. 1.1 — Polyspiral designs.

† The turtle (a small, triangular cursor) lies on a video display. The state of the turtle is completely defined by his location (specified by X and Y coordinates), his heading (with respect to a fixed reference frame), and his pen's being either up or down. The turtle's movement from one location to another with the pen down draws a line on the video display. The Logo commands to the turtle have one or no operands. The no operand commands are 'penup' and 'pendown'. The heading commands are 'right N' and 'left N', where N is the number of degrees through which the turtle should turn. The linear movement commands are 'forward N' and 'back N', where N is now a unit of linear measure (400 of these units measure the width and height of the space within which the turtle moves).

'Squiral' as manifesting emergence in a form investigable in our laboratory environment. This minor insight led to the ruminations in my journal of the following morning, quoted there in full:

variabilized version of my colleague's procedure:

```
  TO DESIGN :distance :angle :decrement
1 FORWARD :distance
2 RIGHT :angle
3 DESIGN (:distance– :decrement) :angle :decrement
```

what I found interesting in her procedure:

	emerging forms:	angle for closed form:	number of spiral arms:
1	triangle	angle = 120	three
2	square	angle = 90	four
3	pentagon	angle = 72	five

For angles near to but not equal to the closed form's angle, emerging spirals form some sort of envelope of the shrinking figure. Note:

1 for zero decrement, the emergent is a circle.
2 for angles greater than the closed form angle, the emergent spiral goes clockwise tighter.
3 if decrement is too big, the illusion of emergence disappears.
the illusion seems to depend on —
a near contiguity of corners in the next ring to the side in any prior ring.
b only a slight precession of the basal form ...

When the general issue of the cause of this illusion of emergent form had arisen, I realized it could be investigated simply by using a higher level procedure to manipulate the DESIGN procedure. The specific question I posed for myself was about the nature of the transformation of a three-armed spiral (seen with angles near 120 degrees). How did the change in arms and the directional sense occur? Was there some sort of flipping phenomenon in the central region? Or degeneration to an amorphous mess? I had no trustworthy intuition on this point, but the computer could come to my aid by displaying those designs at a level of detail which would permit me to see what I could not imagine.

A SUPERPROCEDURE FOR EXPLORING TRANSFORMATIONS: A FAILURE:

I thought the best way to see the transformation from three to four spiral arms would be through an executive procedure which would step down from 120 to 90 degrees, displaying for each angle the appropriate design. Some of the terminals on the Logo system at that time permitted the displaying of 'snapshots' of drawings made by procedures. I intended my executive procedure to display in rapid succession the snapshots generated by a second super-ordinate procedure. When I tested these procedures, I failed of my objective because the software support for the snapshot facility was limited with respect to how many drawing lines could be stored at one time. This impasse exhausted the time and effort I was willing to spend on this idea so I dropped it.

BOUNCING BETWEEN IDEAS

I worried about Marvin's dictum that one should be concerned with 'evolutionary genesis'. I vaguely recalled having read in Susanne Langer's *Philosophy in a New Key* (Langer, 1948) some speculations about the genesis of language. I re-read her recounting of the 'Festal origins' hypothesis. That book was written in the early 1940s. I turned to her more recent work *Mind: An Essay on Feeling* (Langer, 1967), looking there for new ideas she or others had on the issue of how language came into being — and read more of that book than her references to that single issue. How does such a divagation connect with the geometric designs of our alternate theme?

We had run into a technical impasse; here is how it was resolved. Explaining to a visitor at the lab that one of the values of a computer system lies in an expansion of one's intuition, I showed him my colleague's DESIGN procedure. Explaining the use of XORDOWN, a command available on a second terminal type attached to the Logo system, I stumbled across two facts that I had already known: first, designs on these second terminal types (TV raster displays with bit-map representation) are not limited in complexity by being restricted to a small list of lines as were the other terminals; secondly, the drawing speed of the TV terminal types could be fast enough to permit my program for exploring emergence to carry on — if the drawings were made with the turtle 'hidden'. Once this realization was achieved, I coded an executive procedure to manipulate the DESIGN procedure.

THE RESOLUTION IN DETAIL

Reading over the next few days in Langer's book *Mind*, I encountered the drawing reproduced as Fig. 1.2(a), her example of an emergent form dominating its component figures. My reaction: 'I can do that easily, and doing so would add a second level of emergent form to the exploratory procedure I am developing'. I encoded a new executive procedure, SPADES, below:†

```
  TO SPADES :distance :angle
1 CLEARSCREEN PENUP HIDETURTLE
2 RIGHT 90 BACK :distance
```

† I do not expect anyone to interpret this procedure in detail; I present it as an example of a very limited, fixed-purpose procedure. A drawing made by the procedure is shown in Fig. 1.2(b).

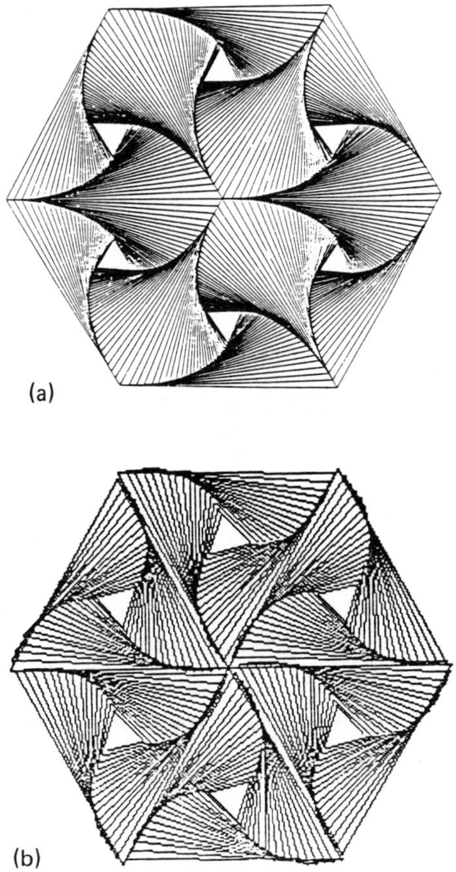

Fig. 1.2 — Emerging forms. (a) Drawing from Langer's book *Mind* (Langer, 1967). (b) Drawing made by the SPADES program.

```
 3 LEFT 90 PENDOWN
 4 MAKE "COUNTER 6
10 MAKE "BASE HERE
20 RIGHT 30
30 DESIGN :distance :angle
40 PENUP SETTURTLE :BASE
50 RIGHT 30
60 FORWARD :distance PENDOWN
70 RIGHT 30
80 MAKE "COUNTER (:COUNTER–1)
90 IF EQUAL ZERO :COUNTER STOP ELSE GO 10
   END
```

This procedure is rigid, drawing six figures in emulation of Langer's design regardless of how ill they might fit together when the component shapes are not triangles. In a quickly made modification, I composed a final version of a top-level control procedure, renamed MPOLYSPI (multiple polygonal spiral) and cleaned up for public use. This 'cleaning up' involved an analysis of what variables might possibly be signficant in executing the designs and permitting their manipultion at the top level of control. With different settings of the variable values, such various designs as those of Fig. 1.3 are generated. This brought the development to a natural end.

REFLECTIONS

After being sensitized to the importance of the interplay of emergence and control, the pattern in the development of this procedure we observe is as follows:

(1) a special purpose construct is made;
(2) that construct is variabilized to remove accidental constraints;
(3) a super-ordinate level of control is introduced to manipulate the most interesting variables of the lower-level components.
(4) the final product is rationalized for public accessibility.

One can name the first step a concrete inception, the second a generalization, and the last a consolidation. Step (3) has no name, yet it appears to be the most significant step in the development: first, because it presupposes selection of some range of variables as possibly the most interesting dimensions in the lower level structures; secondly, because it is the essential step in the development of structure. Call it *elevation*. Notice that elevation is a concrete construction based on a survey of the generality of the lower level structures. It is not a mystery in any sense at all.

One may fairly ask whether this pattern is idiosyncratic or whether it may be characteristic of programmers. The following observation by professional mathematicians, writing under the name of Nicholas Bourbaki, argues that the proces of elevation is more general. I believe it may be a significant component in what Piaget called reflexive abstraction (Piaget, 1971, p. 320):

> A mathematician who tries to carry out a proof thinks of a well-defined mathematical object, which he is studying just at this moment. If he now believes that he has found a proof, he notices then, as he carefully examines all the sequences of inference, that only very few of the special properties in the object at issue have really played any signficant role in the proof. It is consequently possible to carry out the same proof also for other objects possessing only those properties which had to be used. Here lies the simple idea of the axiomatic method: instead of explaining which objects should be examined, one has to specify only the properties of the

objects which are to be used. These properties are placed as axioms at the start. It is no longer necessary to explain what the objects that should be studied really are.... N. Bourbaki. (Fang, 1970, p. 69).

GENERAL CONCLUSIONS ABOUT THE GENESIS OF MPOLYSPI

This mildly creative innovation was guided by an insistent concern where both the personal and environmental means were available to construct a theory-in-action for exploring the issue at the center of that concern. The idea begins with a concrete embodiment, the program SPADES. Once this concrete end-product is generalized, the former 'end' is available as a 'means', a procedural tool useful to a super-ordinate level of control.† The elevation of control creates structure. We conclude with the designs of Fig. 1.3, seeing there a flowering of effects from creation of the new level of structure.

Could such 'elevation of control' be implicated in the growth of cognitive structure, generally? This issue is explored in subsequent work reported here and elsewhere (see Chapter 4 in this volume and Lawler (1985)).

† The description of the process in these terms suggests its compatibility with the description of 'bricolage' developed by Lévi-Strauss (1966) and discussed as a metaphor for the processes of mental self-construction in Chapter 1 of Lawler (1985).

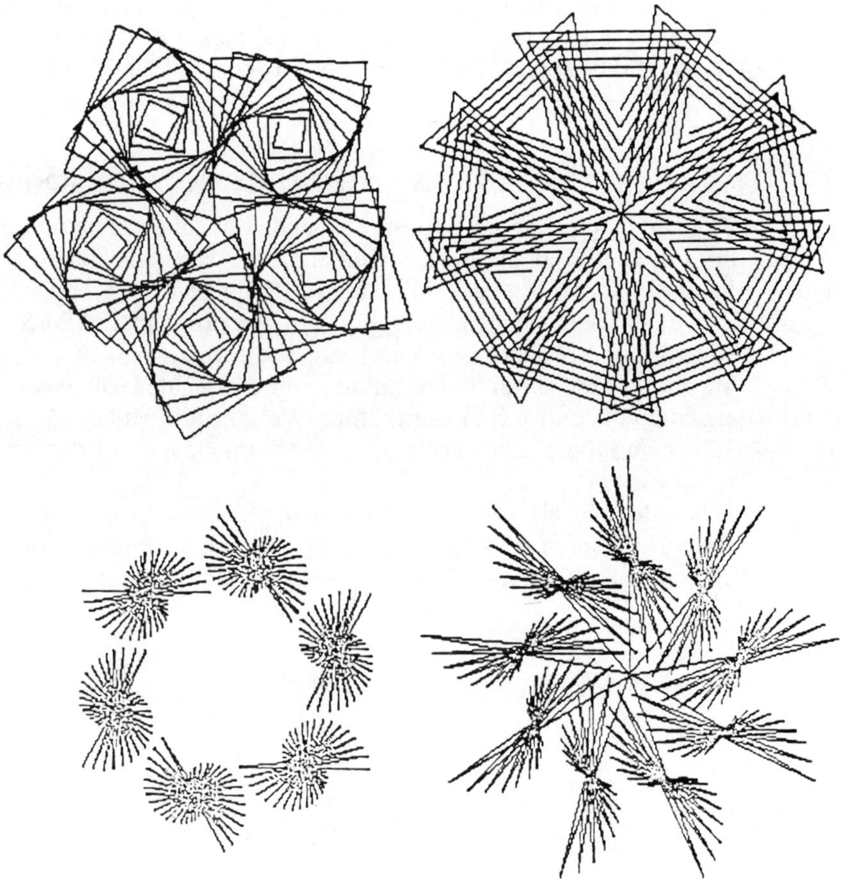

Fig. 1.3 — A flowering of effects. Designs made with MPOLYSPI.

2

Three encounters with number

The encounters described here focus on approaches to teaching elementary mathematics. As a college student in the 1950s and later as a computer systems engineer I had met the set theoretical formulation of the calculus and the need to do arithmetic in variously based number systems. The 'New Math' was not new to me when I met it in my son's kindergarten and first grade classes. Issues surrounding the use of set theory as a basis of elementary mathematics mark the first encounter with number.

Entering graduate school at MIT after a decade in the commercial world, I developed some computer-based tools for elementary mathematics instruction. The next two encounters with number relate my son's experience with those programs. The concluding discussion asks how studies such as this should affect our objectives for math education.

A FIRST ENCOUNTER

Imagine this scene: a kindergarten class, with the teacher and six children standing before the others, three girls on one side and three boys on the other. The teacher instructs the five-year-olds: 'This is a set of three boys, and this is a set of three girls. The number of this set is three. [Pointing at the alternate group now.] What is the number of this set?... Three is the number of the set of three girls. When we make a union of the set of three boys and the set of three girls, we have a set of six children. What is the number of the set of six children?...'. This unfortunate woman, an exemplary kindergarten teacher in other circumstances, was a captive of the content. She told me later she thought the material was dreadful; it was impossible to connect it in any but the most superficial way to ideas meaningful to the children. When one day I joined my son's first grade class, the lesson of the day was on the intersection of sets.†

† For example, a pea is in the set of 'all green things' and more or less, in the set of 'all round things'; we say these two sets, 'all green things' and 'all round things', intersect because they have in common at least one member, as exemplified by a pea. That's all there is to it.

The following weekend, I asked my son Robby why he studied math at school. After a very long pause, he replied, 'Well... I guess Mrs. Meanswell likes to teach math'. Even though this teacher had the best intentions, she was trapped by the curriculum and could not shape it to the children's interests. When a few weeks later Rob decided to quite school because he already knew how to read and didn't want to learn any more math, it was clear I had a problem. In first grade he had already concluded that school was a bore and to keep your mind alive you had to tune out the instruction.† If my evaluation of this situation may seem a bit too engaged for objectivity, listen instead to Marvin Minsky, mathematician, computer scientist, and student of human intelligence, in his criticism of the 'New Math':‡

> By the 'New Math' I mean certain primary school attempts to imitate the formalistic methods of professional mathematicians. Precipitously adopted by many schools, in the wake of broad new concerns with early education, I think the approach is generally bad because of form–content displacements of several kinds...
>
> Historically, the 'set' approach used in the New Math comes from a formalist attempt to derive the intuitive properties of the continuum from a nearly-finite set-theory. They partly succeeded in this stunt... but in a manner so complex that one cannot talk seriously about the real numbers until well into High School, if one follows this model. The ideas of Topology are deferred until much later. But children, in their sixth year, already have well-developed geometric and topological ideas; only they have little ability to manipulate abstract symbols and definitions. We should build out from the child's strong points, instead of undermining him by attempting to replace what he has by structures he cannot yet handle...
>
> The set theory is not as the logicians and publishers would have it, the only and true foundation of mathematics; it is a viewpoint that is pretty good for investigating the transfinite, but undistinguished for comprehending the real numbers, and quite substandard for learning about arithmetic, algebra and geometry... .

The set theory is 'quite substandard for learning about arithmetic'.

The world famous Swiss psychologist and epistemologist Jean Piaget cautions didacticians about children's development of mathematical concepts thus:

† My sympathy with Robby was that of fellow-suffer as much as father. Despite my entering Caltech as a freshman with a perfect score on the ETS Advanced Math Test, I had difficulty with the set theoretic formulation of calculus that was then being introduced to college freshmen. This new math had gone from college material to public school kindergarten fare with no increase in success. I believe an abstract approach to mathematics was a disservice to me as a student and has been so to many another besides. Yet I was eighteen and capable of formal thought. What a worse disservice to subject children, whose mental capabilities have yet to master formal operations, to a mass of incomprehensive abstractions.

‡ The following excerpts are all drawn from a more extensive discussion in 'Form and Content in Computer Science' (Minsky, 1969).

> It is a great mistake to suppose that a child acquires the notion of number and other mathematical concepts just from teaching. On the contrary, to a remarkable degree he develops them himself, independently and spontaneously. When adults try to impose mathematical concepts on a child prematurely, his learning is merely verbal; true understanding of them comes only with his mental growth (Piaget, 1953).

Richard Feynman, Nobel laureate in physics and an outstanding teacher, observes that practically everyone uses arithmetic, that that hardly anyone uses set theory as do the 'pure mathematicians', that there is a growing class of users of branches of mathematics of a high form (for example, engineers, business men), and that we should provide an early mathematical training that will encourage the type of thinking such people will later find most useful. In Feynman's words (1965):

> What we have been doing in the past is teaching one fixed way to do arithmetic problems, instead of teaching flexibility of mind — the various possible ways of writing down a problem, the possible ways of thinking about it, and the possible ways of getting at the problem... .
> The main change that is required is to remove the rigidity of thought found in the older arithmetic books... .
> Mathematical thinking... is a free, intuitive business, and we wish to maintain that spirit in the introduction of children to arithmetic from the very earliest time... .

I judge these goals Feynman proposes as essential: if we aim to help children be flexible in solving small problems today, it is partly in the hope that they will become productive and flexible citizens tomorrow. But how, specifically, can one hope to approach the goal of introducing mathematics in such a way as to foster the child's flexibility of mind? And to do so even at the earliest stage, when children do not have a well developed concept of number. One method, proposed by Seymour Papert,† is to provide even small children 'a mathematical experience more like an engineer's than like a bookkeeper's'. The next two encounters I report tell of my son's use of programs in our research laboratory, the Children's Learning Lab, at the MIT Logo project.

A SECOND ENCOUNTER

ZOOM is a computer program which even pre-reading children may execute. My intention with ZOOM was to provide children with a engaging experience, wherein numbers would be seen, gradually, to play an import-

† The founder of the MIT Logo Project. Papert's early vision was set out in three papers. See Papert (1971a, 1971b, and 1973).

ant role through being useful to the child in meeting his own objectives. When the child presses a key on the computer terminal, the program generates a command in the Logo computer language to drive an output device, which we call a 'turtle'. ZOOM, proposed for use by pre-readers and novices visting our project, was an interface permitting a newcomer to run the turtle very simply: keying a single letter, F, makes the turtle go forwards a fixed number of units; an R or L makes the turtle turn through a different fixed number of degrees; U and D set the pen up or down. This simplification for the novice presents a uniform image on the child's first encounter with the computer.† Even more, he can command actions without first having to comprehend the measures of the domain; this is especially important for those who may have in this system their first encounter with rotational measure, e.g. degress. But most importantly, the child need not confront numbers at all until he himself generates a goal or objective which he is unable to achieve using the pre-set, default values of the operands. The child develops his own need-to-know before having to confront a new piece of knowledge.

Suppose, for example, the default value of rotation is set at 90 degrees, and the child wants to turn through a smaller angle. The child can modify the default rotation, but doing so requires his confronting numbers as a measure of action. For a pre-reader, this is the teacher's opportunity to explore the child's understanding of number and to work out with the child the way numbers are applied in this turtle world. The common technique in our lab called 'playing turtle' has the child and teacher, away from the computer, take turns pretending to be the turtle and acting out the directions the others gives. Thus the child has a chance to connect his knowledge of himself, his own body and its movement, with the new knowledge he is learning in the computer drawing world. He can ask and answer for himself the ways in which he and the turtle are similar and different. Let me recount now some details of Robby's encounter with number through turtle geometry at the Children's Learning Lab.

When Robby (aged 6 years 4 months) first entered the Children's Learning Lab, he saw a logo-person about twice his age executing the Lunar Lander program. The visually striking aspects of that program were its drawing of a moonscape (with cliffs, a building, a canyon, etc.) and the use of the display turtle to represent the landing vehicle. Robby decided he wanted to make one of those. Upon starting the terminal session, I explained that we could run a program called ZOOM that would make it easy for him to draw pictures. We printed out the instructions; when I read them to him, Robby didn't listen. We practiced a little with the basic commands, F [forward 20] and R [right turn 90 degrees], cleared the screen, and then Robby undertook drawing a moonscape.

But there was a problem right away: Robby wanted to start his picture at the left side of the screen; the turtle returns to center screen when the screen is cleared, and whenever it was moved an unwanted line appeared in the

† More specifically, each action is represented by a single keystroke. Further notes on such single key interfaces appear in Lawler (1982).

display. In solving this problem, Robby and I went to look at floor turtle.† On the floor turtle we looked at the pen mechanism, made the pen go up and down, made the turtle draw and then move without drawing. With this concrete example of the stae of the pen, I was able to explain that the display turtle also had such a pen, which you could not see, but whose effect was the same.

The process this example exhibits is the following: a frustrated objective becomes a particular problem; solving that particular problem involves finding the most concrete form of the mechanism involved and showing how that mechanism works; finally, the situation in which the problem occurred is re-described in terms which permit the child to connect what he knows and sees with the concrete problem situation. Robby solved the initialization problem and began drawing quite happily from the left edge of the screen. He made a cliff, a landing site, a building... and developed a problem: the width of his building (20 units) was the smallest unit of distance he could command the turtle to move, so he could not make any windows. He let the problem slip by and went on to draw another cliff going down to a canyon. When Rob wanted to make some tiny rocks at the bottom of the canyon, his frustration became very clear. He asked me how to do it. I warned Robby that the explanation would be a long one and asked if he really wanted to know. He did.

Every child has a concrete mechanism for movement, his own body. We played turtle: 'Forward', I said. Robby took a step. 'Forward': another step. 'Forward', I said, 'two steps'. 'Do you have to tell the turtle how many steps to take?' We went to a different terminal with the floor turtle. I keyed 'forward 1' and the turtle twitched. 'Did he move?' 'Yes, Dad, but he didn't go anywhere'. I keyed 'forward 100'. Robby came to the terminal and examined the printout. He noted that 100 steps didn't take the turtle very far. The conclusion was clear that turtle steps were very tiny. Back at the display running under ZOOM, I explained to Robby that I had told the turtle to take 20 turtle steps every time the 'F' made him go forward. I asked Robby how far he could go in 20 steps. Robby paced off 20 steps and noted, from the hallway, that turtle steps are tiny. After I showed him how to alter the number of steps the turtle would move on a forward command, Robby proceeded to make a few rocks.

Rob then became puzzled again: he wanted to draw a line on a slope, but the turtle always turned square corners. How do you explain to a child so young what degrees are? Robby had seen the turtle turn through 90 degrees when he keyed a single R or L. We played turtle again, both making 90-degree turns when we said 'right' or 'left'. I told him those square corners were 90-degree turns and I said when I turn right or left, that's 'how much of a turn' that matters to me... but the turtle is different. Much as the turtle takes very tiny turtle steps, it makes very tiny turns, called degrees, much smaller than our turns. The number 90 meant very little to Robby. He could

† A device controllable by the same commands which moves and draws on the floor instead of on a video display.

say the name and count that high, but this was the first time that he had seen it made special by being the number that gets you square corners when you want them. He asked what was a good number to use for making sloping lines. I told him 15 degrees. The number 15 is also special in rotation, for it's a common divisor of 30-, 45-, 60-, and 90-degree angles. The idea that there are domains with 'good numbers' was a definite enhancement of his previous concept of number.

This session continued for over two hours at Robby's request. He finished his moonscape and was justified in feeling pround of himself. Subsequent sessions with ZOOM were interesting to him, but the content related less to number and more to ideas of controlling your procedures and planning your work and so will not be discussed here. The final number-related point that surfaces in the use of ZOOM (one which re-appears in the discussion of ADDVISOR) is the question of what a 'correct answer' is. Recall that 15 is a 'good number' of degrees for turning. There is no sense in which 15 can generally be considered a 'correct' number of degrees to turn. When the child's own objective is achieved or missed, he can see that a 'correct answer' depends on the domain and his goals. It is important to learn that a 'correct answer' is not always the same thing as saying something that makes your teacher smile.

SUMMARY

The central points of Robby's encounter with ZOOM I now find of interest are these:

- Robby saw his introduction to the computer as providing a new medium for an activity (drawing) he already much enjoyed.
- He came to the computer because it was involved with my work. This fact gave the experience a coloring of social seriousness.
- He developed an immediate objective — drawing a picture of a lunar landscape — by imitating the work of a more grownup kid, as well as he could comprehend that work.
- Because the ZOOM program is a tool whose adjustment is made by changing numbers, his own goals entrained him in problems wherein he had to augment his understanding of numbers.
- He was working independently in the sense that even though I had useful general knowledge, he alone imagined what he was doing and he had to choose those numbers that would solve his problem.

I consider Robby's encounter with numbers through ZOOM the most successful of the three I describe, and there are good reasons it was so. I was with him as an individual tutor. The milieu was both personally and socially supportive. Robby was executing a skill of a kind he enjoyed (drawing), and he saw that as an appropriate activity for a child in that place He found the experience sufficiently engaging that on his first encounter he worked at his project for about two hours of his own volition. I was pleased because this experience showed him that numbers are things you can use, not merely things to talk about.

A THIRD ENCOUNTER

Some few months after being introduced to computers with the use of ZOOM, Robby underwent a considerably different experience with numbers, one more didactic and calculation-oriented than his use of numbers to control a drawing program. Since the program, named ADDVISOR, derived from ideas on left-right distinction appreciation and the kinds of errors children make in learning to add, I will describe the intellectual framework of the programs first and then present the observations on Robby's encounter with it.

The framework of ideas

Teaching children reading is considered very important in our society. First graders spend hours each day at reading; and we teach them to read from left to right. When they approach the task of bi-columnar addition (usually for a short duration every day near the middle or end of second grade) we instruct them to proceed from right to left. Understanding the explanation of this change of direction, if such is ever offered, requires the child's prior comprehension of addition in a place-value number representation.

Adding right to left is especially hard to understand when the most important numbers are always on the left. Where precision is not required, adults frequently round and estimate by the sum of the higher order digits. Some people retain their preference for adding from left to right. An example: at a nearby pastry store, I purchased two diffrent cakes; one kind cost 54 cents and the other 32 cents. The saleswoman, a woman in her forties or fifties, calculated my bill thus:

```
50–4
30–2
----
80
 6
----
86
```

The woman's use of dashes indicated, of course, separation, not subtraction. This procedure is adequate for everyday use but becomes very cumbersome if you want to add larger numbers, for example, 7438 and 5753.

ADDVISOR

During the course of approximately six hour's work, some of it with ADDVISOR, Robby progressed from single-digit sums to adding 7438 plus 5753. He understood that one could add either from the left or the right depending on his choice in various circumstances. He was also able to indicate why right addition with carries is preferred to left addition for large numbers.

ADDVISOR is a program designed to permit contrast between three methods for adding two two-digit numbers (computer scientists call these different methods 'algorithms'). The contrast is made visible by drawing the numbers and signs on a video display scope connected to a computer (the video display screen of the turtle world). At any time, one or two *contrasting forms of addition are displayed on the screen simultaneously*. For example, adding left to right may be shown on one half of the screen while adding right to left with carries is shown for contrast on the other half of the screen. ADDVISOR controls, flexibly at the direction of the child, the sequence of algorithm steps. The ADDVISOR program performs NO addition at all; should an incorrect sum be entered by a child, that sum will be displayed and the program will be none the wiser.

Place value is a central concept in our representation of the natural numbers. Children encounter this fact as a problem for the first time in bi-columnar additions for those cases where the units sum exceeds 9. Many children add from left to right; most, I aver, proceed initally on the assumption that the columns of numbers can be added independently of each other. Contrast these sums:

I	II	III	IV
20	23	24	24
+20	+26	+26	+26
40	49	410	50

In sum III, the columns of numbers previously manipulated adequately with the assumption of independence are now seen to interact. Advancing from sum III to sum IV may be achieved in two ways. First, the child may accept as true, on expert authority, that following the standard algorithm, which he can be taught, produces a correct result. A better goal is to render explicit and accessible to the novice arithmetician the idea of place value and the interdependence of columns in vertical addition. Such is the design objective of ADDVISOR.

That objective is approached by visibly focussing on the elements that are involved in each step of whatever addition algorithm is chosen by the child. That focus is achieved by placing a box around the digits to be added and the places where the partial sum must go. For examples, see Fig. 2.1. The visual display manipulation capability of the computer makes possible the easy control of the placement and removal of these boxes in preset locations in a sequence determined by the selected algorithm. This implementation provides a crisp and non-confusing focus on the appropriate

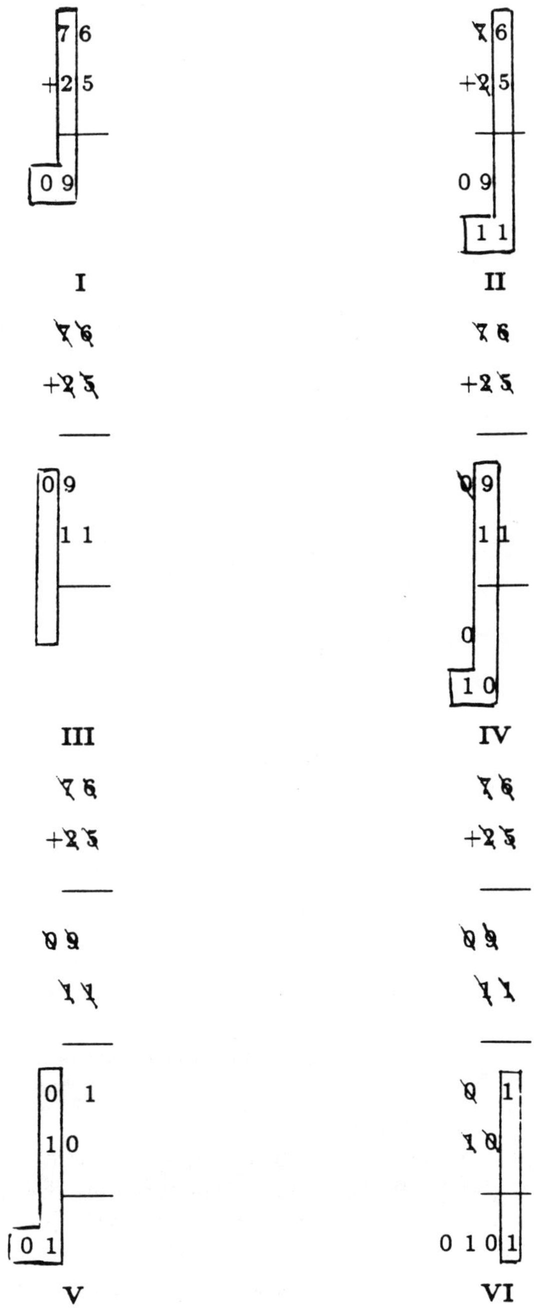

Fig. 2.1(a) — Adding from left to right.

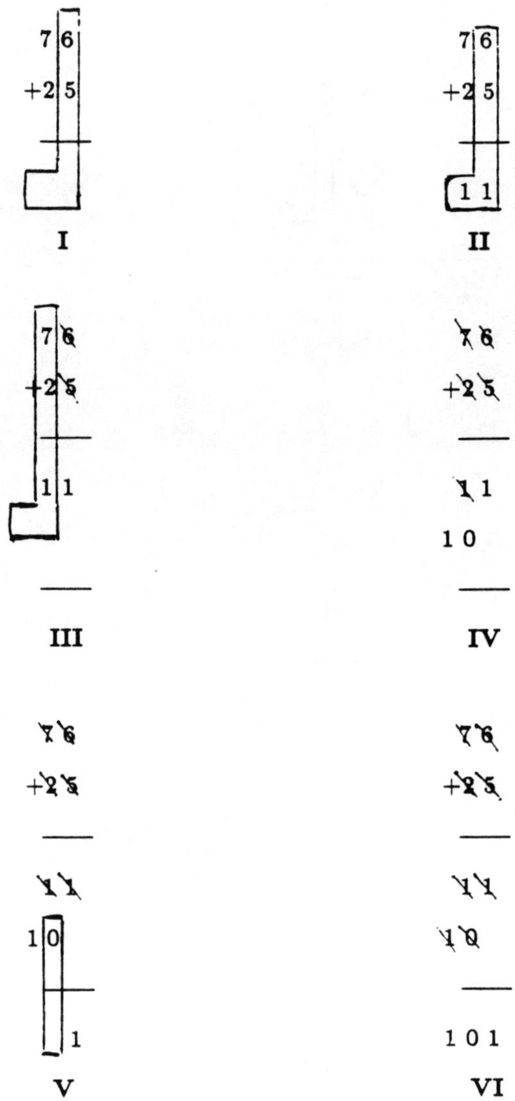

Fig. 2.1(b) — Adding from right to left.

single-digit sums. With no intruding corrections of occasional single-digit errors, and with the sequence of operations visibly and rigidly defined, the child is able to elevate his perspective on the process of adding. He can see and contrast the processes as the addition advances from left to right or right to left at a pace he controls by advancing one addition or the other one step at a time.

The implementation uses these means to place the maximum of control in the child's hands. He can pick out whatever numbers he wants to add between 1 and 99. In practice, since children don't see much reason for choosing one number rather than another, they will be quite flexible in negotiating with some human adviser the numbers to be added. The second important fact is that the child himself does all the addition. Who wants a machine to tell him he made a mistake? As a child comes to see the power of even this arithmetic, he will want to require of himself precision in low-level computations so that he may depend upon his result. Parents and teachers should not require that precision as a prerequisite for learning a different type of knowledge.

The child's step-wise control of the addition algorithms is represented in Fig. 2.1. In the *left-to-right* addition procedure boxes are placed around each single-digit addend pair in its turn. All those sums which are the sum of two digits have the shape of a reversed 'L'. This is a formal requirement, for the sum need not have two digits. Those sums which derive from a single digit (the bringing down of a digit) are enclosed in columnar boxes, for there is no case in which a single digit will generate two in a sum. These simple, formal arguments are demonstrable to a child. Once they are understood, the child can then follow or, as he gains confidence, predict the course the addition algorithm will take on the basis of these local constraints and his earlier decision to add from left to right or right to left. Contrasting the working out of the various algorithms, step by step, makes clear the relative complexity of the three processes. To be able to specify what to do next and to know the consequences of a global decision (which addition procedure will I choose?) constitute understanding a process. You may understand a process even if you make occasional errors in its application. The working out of these algorithms is more like a demonstration than a telling. The computer is used not so much as an 'aid' to the student as a medium of construction in which flexible examples can be built to serve as a local domain for exploration by the child.

The execution of ADDVISOR provides examples of three procedures whose sequence is determined by the formal characteristics of the data and not by the variable values of the data items. It further permits a formal definition of an 'answer': an answer is the content derived from applying a procedure over and over till you can't apply it anymore. It is the data content of a state, reached by the reiteration of a procedure, which prohibits application of the procedure. This sounds like fancy talk; more simply, as a child would see it in Fig. 2.1, an answe is when you have all the numbers in a line. Why do we usually add from right to left? A direct answer is that right-to-left addition with carries is the simplest algorithm of tolerable internal complexity. Rephrased: it lets us put our numbers on one line right away. ADDVISOR exhibits these ideas in a form accessible to first graders. Let us now proceed to recount Robby's experience with ADDVISOR.

At the age of six years and seven months, Robby visited the Children's Learning Lab over a period of four successive days. His arithmetic education in the public schools had progressed uncertainly through most of the single-digit sums; as a secondary procedure, for use when he could not recall a sum, he would resort to representing the individual numbers as hash marks and counting the total to determine the sum. He inclined to estimate the sum of two two-digit addends as ten times the sum of the leftmost digits. (He said later that this idea was his own invention). I asked if anyone had ever pointed out to him how the sum you reached in one column of numbers might affect the sum in another. He said that no one had done so. Whether his recollection was accurate or not, it was clear that he had not been able to absorb and comprehend that fact. I will report Robby's encounters with ADDVISOR as seven episodes (four involving use of the computer and three not so). For convenience, I will label them by the day and time of day of their occurrence.

ADDING WITH ADDVISOR

Sunday evening

On the drive to Cambridge, I told Robby I had written a program to help him learn about adding. At Logo, we started up the computer and read in the programs. Robby chose to add 50 and 65. He could read the words 'LEFT' and 'RIGHT' displayed above the columns of the addends and understood when I told him he could start adding either with the left or right column of numbers. He chose to add left to right. Pushing start caused the display of a reversed 'L'-shaped box around the digits of the tens column of the numbers (see Fig. 2.1(a) for an example). I explained the purpose of the box: first, to show the digits to be added next; secondly, to make a place for the sum of those two digits. With a selected example, on scratch paper, Robby showed himself how adding two digits could sometimes lead to a two-digit sum but that sometimes the sum was only one digit. At this point, I used ADDVISOR's ignorance of how to add single-digit sums as an explanation of why it always allocates two digit positions for the result of a two-digit sum.†

After he was satisified that he understood why the L-block was used, Robby pushed start, thus directing the computer to 'do the next thing'. The computer printed on the terminal log '5 + 6 = ?' Responding '11' and carriage return posted his sum to the display screen in the allocated open space of the L-block and put slashes through the numbers 5 and 6. Pressing start and 'doing the next thing' removed the L-block from the screen. These three steps, then, comprise a complete column addition: box placement, digit addition, box removal.

Upon completing the tens column addition, I asked what we would add next. Clearly, we would add the units column. Well, then, where will the box go? Since the addition result might have two digits, ADDVISOR must make room for two digits while avoiding overlaying the rightmost digit of the intermediate sum from the tens column (see the placement Fig. 2.1(b)). When the units column addition was completed, I asked whether we had an answer, whether or not we were finished. The visible test is that the sum for all the columns are on the same line. Following such steps and arguments we continued till we reached an answer after three addition passes from left to right.

† In this case, the unnecessary stupidity of a computer-based tool is the justification for making an educational point.

Since the units column sum of 50 and 65 was less than 10, no inter-column interaction occurred. I asked Robby if we could add from left to right again. He agreed and let me pick two numbers, 76 and 25. While re-initializing the program, I asked Robby if he coud figure out what the answer might be. His guess was that it must be in the nineties because 7 plus 2 equals 9. He thereupon wrote down the addends and proceeded, as in sum III above, to produce the sum 911. He noted that it could not be correct because 900 is 'way too big.' Oh, it's 91. From this state of uncomfortable speculation we proceeded to adding from right to left.

The unit's column addition proceeded with no complication. However, when I asked what sort of block would be applied to the tens column and where it would go, Robby said, 'Oh, it's impossible. I don't want to do it.' This was the situation.

```
   76        76
 + 25      + 25
 ----      ----
   11        9
              11
           ----
            101
  R–L       L–R
```

We started up an addition on the second side of the screen and performed addition using the same numbers (76 and 25) from left to right. We worked through the three addition passes from the left and reached the sum '101'. Robby expressed surprise that the sum was not in the nineties. When I asked why it was not, he looked back at the display and said that you had to add the '1' from the '11' and that 9 and 1 made 10.

It was hard to interest Rob in returning to the first side of the screen. Once I forced the issue by going on, he became caught up in the details of the process, predicting where the next box would go, whether it would be L-shaped or columnar, performing the additions of the single digits. Robby was quite surprised again on coming to the end when he realized 'it's going to be the same'. *He apparently had made the natural assumption that if you manipulate numbers differently you get different answers.* Having contrasted the adding of the same set of numbers, we were able to see one result and judge the relative difficulty of the two procedures. Robby, although admitting that left to right addition was harder, said he preferred that method at the end of our Sunday session.

Monday morning
We negotiated the numbers for adding. Robby picked 24 and 25 but willingly accepted 27 as a substitute for 24. He either failed to see the coming complications or did not care about them. He freely chose to begin adding

from the right but would not venture a guess as to what decade the answer would fall in.

Robby worked directly through to the sum 52. Before we began adding the same numbers from left to right, he predicted the answer would be different. Thereafter, with both forms before him, he understood in this case why the answers were the same. I asked him why there was a '4' (the intermediate result of the tens column addition) in the left–right addition while no '4' appeared in the right–left addition. He explained, pointing to the right to left form, 'We've already added in the 1 over here'. Pursuing the point, I asked whether the answers should be the same or different. Robby said, 'They should be different; we did it different ways'. I objected, 'But the numbers are the same, 27 and 25. Should they always add to the same sum?' Robby agreed, 'It has to be the same'. I continued, probing to raise the idea of checking an answer: 'What does it mean if you did it different ways and got different answers?' Robby responded, 'It has to be a different problem'. Robby had become restive at this point and turned his attention to drawing a steamroller with ZOOM.

Monday afternoon

Some six hours after the morning session, I put the numbers 46 and 27 on the chalkboard in my office and asked Robby if he could perform the addition. I continued with my other work.

Robby started at the left, writing 6 in the tens column (Fig. 2.2). To

Fig. 2.2.

perform the units column addition, he spontaneously wrote an L-shaped box on the chalkboard, then filled in the intermediate result, 13. He continued by erasing the box, drawing a second horizontal line and summed the 6 and 13 to 73. I asked if that was the correct answer. When he vacillated, I asked if he could add the sum the other way. Robby started on the second addition of the same numbers from the right. He used an L-shaped block and placed the intermediate sum 13 in the appropriate place. I erased that L-block for him because the lines were very close to all the digits. He placed another L-block in the tens column, then rewrote the tens column addends as a horizontal equation: $4+2+1=7$. Then he wrote the sum, 7, in the L-block in the hundreds position. I asked, 'Is that right?' No answer. 'Not quite, Robby, put the 7 over there'. Robby replied, 'I don't want to do it anymore'. To my question why, he replied, 'I know what it's going to be, 73'. I asked, 'Is that right?' He answered, 'Yes'.

Monday night

Out to a late dinner with him and a friend, I asked Robby to explain what we had been doing that day. I wrote 48 and 27 on a 3×5 card. Robby added left to right, using a single L-block for guidance in the units column addition. (He found it necessary to count hash marks to sum 7 plus 8; this is not surprising since he had been up and active for fifteen hours at this time.) When I asked if his answer was correct, he said he was pretty certain.

I then re-introduced 76 and 25. Abandoning the previous method, Robby wrote 9 under the tens column and on a lower horizontal line, both digits of the 11 under the unit column. Then in his second-pass addition, with that alignment, his answer came to 911. His perplexity was immense. Robby recalled then the work of the morning and his answer '101'. When asked to show why he thought 101 was the correct answer, Robby started marking down hash marks. He kept going till 76 (more or less). We asked if there was some easier way to show us. Robby then explained that the answer would be in the nineties except that we had to add the one from the eleven and that made the nineties a hundred.

Tuesday

After the demanding day that Monday had been, Robby showed resistance to using ADDVISOR. He agreed to add 16 and 18 from the left and 25 plus 46 from the right. When I asked him which method he preferred, adding from the left or right, he replied 'any one'. Robby refused to work anymore on addition. Failing to engage his interest in the evening session, I held him to the task only long enough to show him one rapid demonstration of adding right to left with carries.

Wednesday morning

Engaging Robby's interest remained a problem. He explained he was mad at me because he wanted to go to the Children's Museum and that would be more fun than doing sums. After agreeing to take him there later, I proposed as candidate addends 97 and 56. Robby balked but agreed to go on with 25 substituted for 56. Although the procedure for adding with carries, which Robby chose to execute first, is shorter than the others, we made slow progress. We dealt explicitly with the fact that only columnar blocks enclosed the column to be added, the fact that lack of space for the second digit, the tens digit of the intermediate result, required use of something called a 'carry'. When Robby keyed '12' as the sum of '7 + 5 = ?", only the 2 appeared on the screen in the units position. Where did the 1 go? I explained that the 1 was used by ADDVISOR, but that Robby could use it to answer the question he now confronted (did he have a carry?), for the carry is the number left over, the digit you don't have room for in the line where the answer will be. Robby said he wanted a carry. We went on together and concluded with a correct answer. Robby wanted to stop and not perform the contrasting right-to-left addition because 'it will be the same'. When I asked if he could prove that, he agreed to humor me. At our conclusion with the same answer, 122, I asked Robby which was the best way to add. He replied 'carries', but when I asked him why, he was non-committal. He said he understood carries now and wanted to stop. With that short exposure, I doubted his understanding and asked if he could add with carries on the chalkboard. 'Sure'.

At the chalkboard

Robby picked the numbers 96 and 25 for adding and wrote them in the vertical form. He placed a columnar block around the units column, wrote a 1 in the units column and a carry over the 9. He went back to review the work on the display screen (I believe to avoid having to figure out the sum of 9 + 2 + carry) and then wrote a 12 on the answer line. I asked if he had a carry there, in the tens column. After some pause, Robby said, 'Yes, but it doesn't matter; we would write it down in the answer anyway... and am I done now?'

I said we should do one more sum with carries and wrote on the chalkboard 9768 plus 5844. Robby laughed and said nobody could do that. I said we could together, that I would write down the answer and carries if he would do the adding. We proceeded directly to our answer. Robby was astounded: we had made a number of 15 thousand. This was clearly at the upper reaches of his imagination of magnitude. He asked if we could do another. I chose 856 plus 376. After we completed this sum together, I said he should work out the next one all on his own: 746 plus 365. He asked for another when done; I selected 857 and 424 (these addends have no carry out of the tens column) which he also added correctly. In all these additions, Robby used columnar blocks to keep his work organized and aligned properly. He was obviously delighted in witnessing the power of what he had learned and went to find a friend, Danny Hillis, another Logo lab member, to tell him of his discovery.

Later in the day, I asked Robby if he remembered the addition that made more than 15 thousand; could it be added starting from the left? He replied that it could and when I asked how long it would take, said, 'Probably a year'.

Conclusions

Two days after the computer sessions at the Children's Learning Lab, I asked Robby to add 7438 and 5753 on a chalkboard while I did some bookwork. When he finished, each column was enclosed in a columnar block, the carries were marked in the standard places, and the answer was correct.

I am convinced Robby experienced a conceptual breakthrough. The evidence is both from performance (getting the answers correct consistently in all algorithms) and from explanation (he knows that there are at least three variations on two-addend adding and in what circumstances he prefers to use one form over another).

To what extent is this experience a special case? Let me first summarize Robby's prior developmental position:

— he was capable of adding single-digit sums, though he was uncertain of results beyond 7 + 7; for larger numbers he used the alternate procedure of counting hash marks.
— he could count beyond one hundred and easily read three-digit numbers; with difficulty, he could read numbers in the thousands.
— he showed a useful sense of magnitude; for example, he could recognize that '911' was the wrong sum for 76 and 25 because if was far different from his estimate of 'in the nineties'. Without this last perception, he would not have recognized the inter-column interaction which is central to our representation of the number system.

I believe that a child with these three capabilties is ripe for the kind of experience Robby had at the Children's Learning Lab, and in addition that his developmental configuration is common, although the age at which it is reached may vary. Consequently, I see nothing special in Robby that would make him specifically suited to having such a conceptual breakthrough.

With regard to the circumstances of this work, there are four factors tha one might examine for significance:

(1) as Robby's father, I have a special influence over him (as his father, it is clear to me how over-estimated this influence may be);
(2) the Logo project is a small sub-culture where numbers are much more important than they are in the world at large;
(3) Robby's encounter with the material was intense and compressed in time; the implementation of the ideas was computer-based.

I believe factor 2 is the critical one, that Logo is a place where numbers have meaning through applicability and are important enough to worry about. You need numbers to make the turtle turn or to move forward. You need numbers to draw on the display screen or to play computer games. Big numbers are as common as little numbers, and big people use numbers a lot. Logo is home ground for the natural numbers and a place where Robby feels at home. His sense of belonging at Logo may derive from my involvement there but it is broadly supported by other people's interst in him and in what he thinks. The intensity and time compression of the experience are an accident of his spending only a few days at the Children's Learning Lab and simply made it harder for him to endure. The method of teaching addition embodied in ADDVISOR is not bound to computers but would be much harder to put into effect without them.

If this encounter, as the one with ZOOM, represents a beginning mathematics education in an ideal environment, why did Robby occasionally show resistance to using the resources of the Children's Learning Lab? I could argue that we're all cranky sometimes and so on, but there's more to it than that. There's a common dilemma with new ideas introduced in school: one can't really appreciate the value of an idea until it's understood; but why would anyone want to bother with an idea unless one can appreciate its value? Social support, such as the community of the Logo research lab offers, is one answer to this problem of education, but such a community is accessible to only some of the people some of the time.

Back in school

After his encounter with ADDVISOR at the Children's Learning Lab, Robby returned to his school. During 'show and tell' the following Monday, he explained to his classmates how to add two three-digit addends (he noted that he only gave two examples because he didn't want to take up all the time). Commended for his skill, Robby was then informed that 'in first grade we never add any numbers with a sum greater than 12'. The message was clear to him. Robby later told me that if I really wanted to help him with math I should teach him to do single-digit additions quick so he could score points for his team of boys in their competition against the girls. In his everyday school world, his development was denigrated and his skill was made to appear to no use whatever.

Discussion

One can marshall arguments in favor of teaching a limited set of single-digit additions in first grade:

— it exemplifies an incremental approach;
— it provides a basis of well known facts from which the child can build up his ideas of addition;
— one must be able to add '8 + 7' before one can add '87 + 78'.

On the other hand, limiting the computation problems one poses to children by the sum '12' exhibits a profound miscomprehension of the character of arithmetic, in this case addition. My contrasting view follows.

One basic process is computing single-digit sums, e.g. $2 + 6$ or $8 + 9$. A second process is integrating results from these computations. The fact that a child, or all children, can make occasional errors in single-digit sums does not restrict the ability to comprehend or execute the second process. Would anyone argue that the error in this sum, $468 + 357 = 823$, is any worse than the error in this, $8 + 7 = 13$?

The central fact of our natural number system is the multi-digit representation of numbers. One might argue against computing sums of value greater than 9 because this might entrain you in explaining why one needs two digits to represent the number 10 (understanding that fact may require sophistication beyond the reach of most six-year-old children; but most seem to accept without too much argument the representational fact). Twelve, as an upper limit, is hard to defend unless seen as the sum of fingers and ears (or some other such absurdity). In contrast, consider this sum:

```
  3 4 6 9 2 7 5
+8 5 3 0 4 1 4
_______________
```

Because there is no column interaction (except at the left where it doesn't matter), any child who can sum the columns can compute the sum correctly. Notice also that the child can add left to right or right to left as he wishes. Why do we radically restrict the computational range of what we teach? It is precisely that range that expresses the power of the hindu-arabic notation. Recall Robby's delight in finding that he could add sums as great as he could read. We can easily teach a few-step, direction-insensitive addition algorithm, thus:

```
  3 4  6  9 2  7  5
+5 5  5  5 5  5  5
____________________
  8 9 11 14 7 12 10
```

Now, one must start at the right and check for 'bugs'. if we have a two-digit sum in any column, that's a 'bug'; and '1' must be added to the next left column. Thus we get:

```
  8   9 11 14  7 12 10
       ↙  ↙      ↙  ↙
+     1  1      1  1
______________________
  8  10  2  4  8  3  0
```

And here, we see, we have to check for 'bugs' again; so our answer is 9,024,830.† This proposal, at the level of introductory arithmetic, admits that humans operate with imperfect procedures, the results of which are subsequently fixed up by correcting errors. Another way of expressing this idea is that one should add starting with the most important numbers and making the best estimate you can while ignoring interactions of terms, then correct those estimates by accounting for the interactions caused by the place-value representation of our number system. The correctness and precision of this computation can be seen as a process of getting closer to the 'correct answer' by successive refinements. A richer approciation of this process notices that it embodies the assumption of linearity — that complex processes can be broken up into separable parts for better understanding.

† This proposal embodies the idea that left-right addition is a recursive procedure, a technical observation made by Howard Peele, Professor at the University of Massachusetts at Amherst, in a discussion of an earlier version of this paper.

This assumption is basic to much scientific reasoning, even of the most sophisticated kind.

The last few comments make it sound as though one must know a lot of mathematics or computer science to help children deal with numbers. Not so; not so at all in this case. The essential idea is doing what comes naturally..., adding from left to right and correcting the flaws of that procedure when they occur. Remember the lady who sold me the pastry. She added from left to right and did it well. Remember Robby's estimate that 76 + 25 must be in the nineties because 7 + 2 equals 9. The former shows the usefulness, the latter holds a key. A mathematician has a strong result if he can place bounds on a computation. For a child who has been working on 5 + 6, to realize that he can bound a seven-digit sum may be an engaging result. For example:

$$\begin{array}{r} 5796527 \\ +\,6614723 \\ \hline \end{array}$$

The sum will be more than (5 + 6) million and less than (6 + 7) million. The sense of power a child can get from controlling, even approximately, large numbers, may inspire him to improve his single-digit computational skills leading him eventually to refine his estimates to perfect results.

GENERAL CONCLUSIONS

The conclusions I draw from the three encounters I've described are of these sorts.

— Even granting that the earlier attempts at curriculum reform have failed does not imply that one should go 'back to the basics' if those 'basics' do no more than constrain the range of computation within which children practice rote memorization. The desirable goal, even in the earliest training, is to foster a flexible, results-oriented approach to calculation.

— A good way to introduce children to numbers is to provide them a medium of expression, within which, as their objectives grow more discriminating, their involvement with numbers must increase.

— One should distinguish between the need to memorize single-digit additions and the child's ability to comprehend and execute the assembly of those single-digit additions to multi-digit sums. There is no reason, as Robby's rapid success with ADDVISOR shows, to demand that children should be capable of the former before confronting the latter.

— A natural way to introduce children to multi-digit addition is through estimating the bounds of impressively large sums; a second step is to introduce left-to-right recursive addition as the basis of primary school computation; the intention of such an approach would be to give children a sense of the power of being able to deal with large numbers and a sense that one discovers 'correct answers' by a process of refining first estimates.

But finally, the rigid reception Rob encountered on his return to school is a simple indication of how much inertia exists. It may be important to take seriously Piaget's assertion that children learn about number by themselves more than they are taught by would-be teachers. What those circumstances of natural learning are like is our next theme.

3

Sketches of natural learning

In a famous passage of the classic *The Sciences of the Artificial,* Herbert Simon (1969, Ch. 2) describes the complexity of the movement of an ant across an irregular bed of sand then concludes:

> An ant, viewed as a behaving system, is quite simple. The apparent complexity of its behavior over time is largely a reflection of the complexity of the environment in which it finds itself.

He then asserts his objective of exploring this hypothesis — but with the word 'man' substituted for 'ant'. Simon, and his colleague Newell, have pursued this vision with vigor and precision, doing their best to represent apparently complex human behaviors with architecturally simple computational models†. Their archetypical analysis is that of a mature person solving some variation of one of Bartlett's cryptarithmetic puzzles; the analyses are exquisitely detailed. Their colleagues have worked for 15 years to extend this vision to a broad spectrum of performance on psychological tasks. Some of us remain unconvinced of the correctness of Simon's vision despite this admittedly impressive corpus of work.

It is possible to imagine that through experience people construct internal models of the world and that the details of these models and their interrelations may come to be more complex and more effective in determining behavior than their external environment. The simplicity of manifest

† See their fundamental work *Human Problem Solving,* 1972, and more recent descriptions by Newell of production systems and their evolution, as in his invited address to the 1985 Cognitive Science Society Conference.

behavior witnessed in the problem-solving protocols analyzed by Newell and Simon may be a consequence of a highly developed self-control of mature subjects — a characteristic of behavior rather different from the volatility of the child from whom every adult grows. This example, from *Midwest and Its Children* (Barker and Wright, 1955), may be taken as representative:

> On June 2, 1949, four year old Margaret Reid spent 28 minutes in the Midwest behavior setting, Home Meal (lunchtime) . . . Her behavior was consistent with the standing behavior pattern of the lunchtime setting. But this is by no means all. Margaret did 42 clearly discriminable and different things on the level of behavior episodes during the 28 minutes. Here are 21 of the 42 actions, just half the total:
>
>> Rejecting lemonade; Recollecting pancakes eaten for breakfast; Cutting tomato; Helping self to noodles; Forecasting Bible School picnic; Challenging little brother to lunch eating race; Appraising combination of lemon juice and milk; Inquiring about Valentine's day; Coping with dropped napkin; Commenting on play of neighbor friend; Playing on words about Bible School picnic; Wiping something out of eye; Reporting little brother's capers; Dunking cookies in cocktail sauce; Telling about imaginary friends; Putting box of Kleenex on bench; Inviting parents to look into stomach; Soliciting mother's opinion on brother's eating; Using spoon as airplane; Chanting 'Bones to Be, Bones to Be'; Reporting on birthday greetings at Bible School . . . .

Barker and Wright then expand one of these incidents — Cutting tomato — into fifteen even more detailed action descriptions. How information processing models of today should approach material exhibiting such volatility is far from clear.

There appears to be a lot going on in the human mind. How that multiplex activity relates to problem solving 'in the wild' (as in the mind of a child) may be critical to understanding natural learning.

The organization of these sketches

This collection of observational material is grouped under three headings:

—Complex minds and homely circumstances.
—The character of natural learning.
—Instruction as invention.

The sketches themselves are drawn from a collection of protocols of my son's behavior either at the MIT Logo Laboratory or elsewhere in my presence.† Much of the material was mechanically recorded, manually transcribed, and collated with its appropriate documentation. I present here the more interesting extracts from that larger and more detailed corpus of material.

I believe this material is relevant to studying learning directly, in that the examples can illuminate the character of learning through problem solving; further, the observations also highlight the methodological problems of studying learning. Finally, the examples of developing new activities and experiences to advance learning, based on the particular experiences of individual children, suggest that this may be the area in which the computer, as medium, may ultimately have its most profound effect on education.

COMPLEX MINDS AND HOMELY CIRCUMSTANCES‡

Guests coming to our house one day inspired my wife to bake a rum cake, a favorite dessert of mine. The children knew that with company around they could take advantage of our playing the role of tolerant parents — and they did. I will mention, but not recount, the noise and the various rambunctious exhibitions before and during dinner and now proceed directly to dessert — for that rum cake had seized the children's imaginations.

We all had a piece. Afterwards, I, with my cigar and glass of port in hand, had to admit that the children might have a second piece of cake. And eventually when, in pajamas, they claimed a third piece of cake, I made the best defense I could, agreeing to this third as the last piece; declaring (after the fact) that each could have three, I argued that they had had pieces number three, number two and one, and that only zero was left, no more. Thus, when finished, off to bed they must go. After a few bites, Robby excitedly countered, 'We can have a million more pieces'. He elaborated by explaining that there were 'a million numbers less than one'. I asked what he meant. 'There's one and zero,' Robby replied, 'then there's zero minus one'. 'And what else?' I queried. 'Zero minus two'. 'That's two', I agreed. 'And minus five, and minus one hundred . . . and minus one million'. This completed his proof.

Robby had been introduced to two particular negative numbers before§. I infer that in the few intervening days (if not at the moment of this little episode itself) Robby had invented for himself the negative integers as symmetric to the positive. His excitement, as we discussed this trivial

† This observational material was collected in 1976 and 1977. At that time, the MIT Logo project was in the vanguard of educational computing research. With the broad diffusion of microcomputers in schools, these observations are now more widely relevant than when originally collected.

‡ From protocol 10, at the age of 7 years, 3 months (7;3).

§ Minus two and minus ten. See the next sketch from protocol 6, which preceded this incident by a few days, and from protocol 9 under the heading 'Instruction as Invention'.

occasion of pieces of cake, showed his engagement with this aspect of the number system.

Family customs and knowledge utility†

If what you learn derives largely from the culture you live in, and if individual differences derive support from variations in the home (that micro-culture of the family), it is important for people to know what learning is like in other family's homes. The anecdote I present here reveals one case of the dynamic interplay between social relations and manipulating numbers.

Counting and punishment

In our home, the children have become skillful at manipulating numbers. Whenever they raise a problem which they might figure out, I try to take the time to work through it rather than simply give them some answer which they must accept on my authority. I look for occasions to involve numbers in what they think about. For example, when we have squabbles so serious that someone must be punished, that punishment usually takes the form of 'go sit in that chair until you count to (some number) (in some non-simple fashion)'. The reason this developed is that, when 'punished', the children always wanted to know the term of their sentence. Counting provided a useful termination for the children before they could tell time. It was easy for me at first when the children had trouble counting to 40. They suffered a formal penalty but were too busy counting to feel bad. As they became more competent at calculations, giving them tasks difficult enough to generate a 'cooling off' period became harder.

Enjoying small jokes, I have occasionally asked them to do very difficult things, e.g. 'count to 37 by threes'. I do not advance this procedure as something to be imitated but rather offer it as evidence that the children found number knowledge very useful in their everyday lives.

Competition

Though I love my kids, I would not in any sense claim my family is exemplary, but it is rich in vitality and sometimes in good humor, despite the children's rivalry which surfaces in nearly everything. My daughter Miriam, aged 5 and one half, on visiting the Logo computer lab required of me that I write a program which would set her addition problems to do. This is not a kind of 'computerized education' of which I approve; but since she asked for it, I wrote her a program to quiz her on the sum of two randomly selected digits.

Robby thereupon required a program presenting some more difficult

† From protocol 6, at age 7;3;1.

problems. So I wrote one challenging him to sum one single digit and one double digit addend. Both children were happy for a while . . ..

Miriam mentioned at dinner, with great pride, that she could count backward from twenty by twos. She demonstrated. When she reached zero, I asked Robby (who had been introduced to one negative number, minus 10 Fahrenheit, as the best temperature for keeping ice cream) what came next. He said, apparently somewhat startled himself at the realization, 'minus two'. I congratulated Miriam on her skill. She, with irritating condescension, said that counting backwards was 'cinchy', and so was adding. Robby countered that adding was easy only because her program used easy numbers. Here, she dealt him a severe blow, revealing that she had been adding using *his* program at Logo. She declared that 'cinchy' also.

Robby had trouble mentally summing such numbers as 7 and 16 (Miriam had succeeded with comparable sums by borrowing my fingers to help her reach the answers she counted up to). The battle was joined. Robby, who had become skillful at right to left adding with carries, confidently proposed on his napkin the sum 642 plus 260. Miriam was put down, but she ignored that fact and dismissed the task. He added the numbers correctly (see Fig. 3.1). Miriam, declaring it her turn, challenged Robby to sum 29 and 100. Because she thought it a hard problem, I infer this task can represent the limit of her skill. Robby wrote down the answer directly then had his revenge: he set her the task of summing 66 and 78. Such is a wicked sum for beginning arithmeticians; not only are there carries, but the answer contains a third digit where each of the addends had only two digits.

Miriam was in trouble, but she charged on. 'Can I borrow some fingers?' So we four at the table each raised our ten fingers and around she went, then again, stopping at 78. Next, restarting, she counted fingers by tens to sixty and then six more. He procedure failed to produce a sum different from either of the numbers she was attempting to add. When I asked her about the sum, Miriam decided to back up to a more trustworthy procedure: she began writing hash marks on a paper napkin. The marks spread across the page; returning back on the same row, she made larger hash marks to discriminate them from the first pass. (I count 59 marks in that row; she may have made up the sum to be 66). Her tenacity was exemplary.

Robby's amusement was unconstrained, as though applying such an uncontrollable technique were the craziest action he had ever witnessed. His reaction amused me, for a mere six months before he had exhibited precisely the same response to a challenge of comparable magnitude (see the section about ADDVISOR in Chapter 2). Although anecdotal, this is striking evidence of the amnesia about prior cognitive structures that were superseded by others.

Miriam abandoned writing hash marks for a listing of the integers in order. This second technique, though more difficult to write out, was clearly superior because one need not start at the beginning again if momentarily confused. WHen Miriam ran out of space on the paper, we solved the problem together. Though Robby wanted to continue, I stopped our exercise at this favorable time for all of us, and we proceeded to enjoy the rest of our now-cold supper.

The urge to instruct: showing a solution†

Both children lay on the floor tonight hoping to induce our new Scotch terrier puppy, Scurry, to lick their ears and jump on them. Scurry is obliging and the house has been a pandemonium. Robby thus easily fell under censure for some small failing, and I told him to stand in the hall and count to 27 by fours. He griped that he could not do such an impossible task. I asserted that he could do it.

Robby: Oh, you mean do a three at the end?
Bob: No, just at the beginning
Robby: Oh . . . 3, 7, 11, 15, 19, 23, 27.

Soon after, Miriam, for whatever reason suffering a like punishment, allowed as how she could count lots of ways, by ones, tens, fives, and twos (this is so). I told her to count by fours.

Miriam: Daddy, I can't.
Bob: To 12.
Miriam: That's impossible.
Bob: Prove it.
Miriam: 4, 8, 13.

So she had her joke on me, and in general good feeling we approached the supper table. At Miriam's request, we listened to Beethoven's 5th symphony as we sat down to eat. Robby told me he had finished multiplication fast in school today, that he was first in the class. I congratulated him, but, not wishing to support the competitive sense he was expressing, I decided to change the subject. I noted that I prefer the 6th symphony to the 5th. Both children then chirped up that they liked the 9th best of all. The way they responded seemed as if they were inventing a game of 'my favorite symphony has a higher number than yours', so I countered that I also liked the seventh, and that 6 and 7 are 13. Robby claimed that he most preferred the 9th and the 8th (I don't believe he had heard the 8th at that time) and Miriam, 'I like 3 and 5 and 7 and 9'.

In an unfair counter-move, I noted that I especially liked Opus 132 (the A minor quartet) and was properly shouted down with the complaint that an Opus was some other number, not a symphony. Miriam closed the dispute:

Miriam: I like them all, and that's a lot.
Bob: How many?
Miriam: (beginning to count . . .)
Robby: 9 plus 8 is 17, plus 7 is . . .
Gretchen: 45 is the answer Rob.
Robby: Hey, I'm trying to think and you're distracting me.

† From protocol 15, at age 7;6.

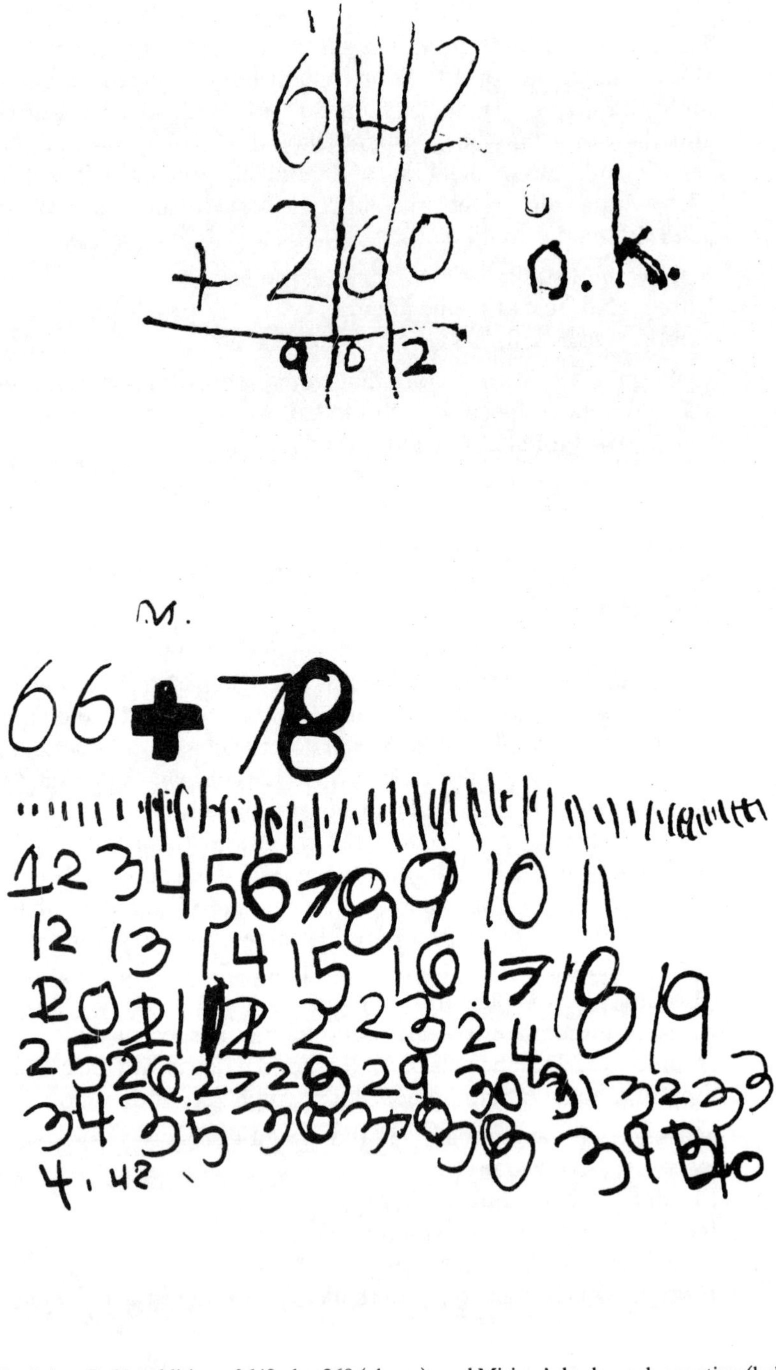

Fig. 3.1 — Rob's addition of 642 plus 260 (above), and Miriam's hash-mark counting (below).

Gretchen then became very insistent and showed Robby the solution to the discovery of which I was trying to lead him:

Gretchen: How much is 9 plus 1?
Robby: 10.
Gretchen: How much is 8 plus 2?
Robby: 10.
Gretchen: How much is 7 plus 3?
Robby: 10.
Gretchen: How much is 6 plus 4?
Bob: (when neither had completed the procedure) . . . and how much is 5?
Robby: 5 . . . Hey, wow! The answer's 45.

Gretchen had no general and pervasive didactic intention, but her aggressive instruction points out how adults who have learned some clever trick insist on showing it to people who have not yet encountered it.

The urge to instruct: emphasizing a problem†

We drive in cars a lot, and I find it hard to be patient at long stoplights. Since my children fidget whenever we wait for a red light, some months ago I introduced them to an old game:

> When you encounter a red light, you chant the formula: 'Red light, turn green before I count to seventeen'— then begin counting. You win if the red light turns green before you reach the number seventeen.

This is a good game because victory is always within your grasp. One may proceed at any rate or with variations of rate he desires.

Miriam, at the age of six, counts initially at her 'normal' pace and begins slowing when she gets to twelve, drawing out both words and intervals. Rob began with the same approach; more recently, in keeping with his developing concern about playing fair at various other games, he has maintained a regular counting pace and thereby 'lost' the game.

As I waited at a traffic light last night, I heard Robby intone: '15, 16, 16 and $\frac{3}{4}$, 16 and $\frac{3}{4}$ of $\frac{3}{4}$' and so on. The light turned green. As we moved out of the intersection, I asked Robby how many three-quarters of three-quarters he needed. He replied, 'Five. But it doesn't matter. I win every time now'. I congratulated him and noted that he had come upon one of the great puzzles of all time. Rob: 'Is it a tough one?' I told him it took over two thousand years for someone to figure out to solve Zeno's puzzles and said we would discuss them more comfortably when I was no longer driving.

The concrete example of the racecourse paradox I gave put Achilles at the end of our driveway with an intention to go to the barn. I argued that before he reaches the barn he needs to go half-way — say to the wall of the courtyard. Robby agreed that he still had halfway to go. And, I continued, to go from there he still must go halfway — and he will have half of a half to

† From protocol 20, at age 7;9;13.

go. And so forth. Robby squirmed a little. 'But if he goes many, many halves of halves, he'll get to where he only has an inch to go'. I agreed, but still maintained that he had not yet reached the barn.

Gretchen added that this was called a paradox, because we all know we can just get up and walk to the barn any time we want. I continued, 'If Achilles has an inch to go, he still has to go halfway, a half inch first — and that still leaves half an inch'. Robby: 'And half a half is a quarter . . . and that still leaves a quarter to go'.

When I asked Rob if he wanted to hear the other paradox about the tortoise and the hare, he answered, 'Not now . . . wait about six years'.

Contrasting the two exemplified forms of 'instruction', the most significant difference seems to be what one is trying to teach. In the first case, Gretchen tried to communicate to Robby a specific procedure that would permit him to calculate sums of uninterrupted integer sequences. In the latter case, I tried to extend Rob's accidental engagement with a concrete problem to introduce him to a broad class of mathematically interesting problems. The former is more like 'giving knowledge'; the latter more like 'setting goals'.

The obvious is unknown†

Multiplication has been an insistent theme for Robby. One night, about five minutes after his going to bed, as I passed through his room, Robby rose out of the covers with a groggy look on his face to ask, 'Dad, is 3 times 30 90?' I answered his question and asked why he wanted to know; he replied, 'I don't know'. At odd moments over the intervening weeks, Robby asked is *N* times 10 equal to some-value. One night at dinner, the following dialogue took place:

Robby: Mom, is 9 times 10 90?
Gretchen: Yes. How much is 10 times 9?
Robby: 9 plus 9 is 18; that's 2. How much is 18 plus 18?
Gretchen: 36.
Robby: 36 plus 4 is 40.

With this pending confusion, I cut off dialogue at this point in the hope that I would later be able to bring up the question of Robby's understanding of commutativity in a clearer context.

Robby apparently used the tens as a discrete scale wherein each decadal increment is 'the next bigger thing'. In an incident from protocol 4,‡ Rob argued that when building a triangle from tinker toys, one used 'the next bigger one' for the hypotenuse; consequently, when drawing a right triangle with an hypotenuse of 50 turtle steps, 40 turtle steps should serve as the

† From protocol 14, near age 7;4; see also the related protocol 13 under the heading 'Guiding constructive analogy'.
‡ Previously reported in Lawler (1985, Ch. 1) at age 7;3.

equal legs because 50 was 'the next bigger' thing from 40. His used of this 'decadal' scale as an estimation basis can be seen in the following incident. After having asked for verification of 3 times and 7 times 11, Robby asked if 11 times 11 was 120. Responding that his answer was not perfect but nearly so, I asked how he estimated the product. He replied, 'Well, I know that 11 tens would be 110, but elevens are a little bigger than tens, so the answer has to be bigger'.

During this period of multiplying by the controlled counting of decades and other number scales, the question of multiplicative commutativity was again raised at table. When Robby showed he knew that 20 times 5 is 100, Gretchen asked him, 'How much is 5 times 20?' To this query, Robby replied: '20, 40, 60 . . . I give up'. The following Sunday, Robby had to face the possibility of missing his favorite TV program, *Victory at Sea*. He complained that he might miss the battle of Midway. Gretchen countered that he would be able to see it sooner or later, that Victory at Sea had been on TV for 20 years and that the battles would roll around again.

Robby: How much is 20 times 52?
Gretchen: 1040 [Gretchen computes accurately and quickly].
Robby: How much is 52 times 20?
Gretchen: 1040.
Robby: [Looking puzzled] It's the same thing . . . How much is 20 times 52?
Gretchen: 1040.

Gretchen avoided telling him about commutativity in multiplying.

Notice that this 'obvious' property, which adults freely use in mental calculation, was something that Robby had yet to discover. In this case, the child's ignorance was of little long-term consequence. But there do exist documented cases where such ignorance of the obvious must severely constrain any understanding we ascribe to a child. For example, Miriam's understanding of strategic play in tic-tac-toe did not recognize the importance of opening advantage (see pp. 131 ff. in Lawler (1985)). In both of these cases, an important aspect of the child's representations is that they must function tolerably well, even though quite partial in the specific sense of omitting significant elements of the mature representation.

CONCRETE ACTION IN A NON-SQUARE WORLD†

Two of Robby's current interests have come together in a productive way. Following my buying him a new tool box, he elected to take up woodworking in an after-school program. His delight in building ship models led him to declare he was going to build a 9-foot long wooden aircraft carrier. I voiced

† From protocol 16, at age 7;7.

absolute objection but, content to let the fantasy grow, said he could plan the ship and scale down the model later to some reasonable size. Later he asked advice:

Robby: Dad, how much is 9 feet take away four?
Bob: 5 feet.
Robby: No, I mean take away 4 inches.
Bob: Come here and I'll show you.
Robby came to my desk and worked out this exercise:

9′	➔	8′ 12″
−4″	➔	−0′ 4″
??	➔	8′ 8″

This dimension, 8′ 8″, is for the sub-structure of the carrier. The 9′ flight deck will overhang 2 inches at each end.

In response to this and other expressions of the need to draw plans, I introduced Robby to orthographic projection (front, side, and top views). Thus when last week I entrained him in my task of building a run for our dog (of 2 × 3s and hardware cloth), it seemed appropriate to ask him to draw plans. This was even more the case because I had formerly asked Robby to compute dimensions on the sizing drawing to determine how long should be the 2 by 3s we needed to make the frame. We were to enclose the two ends of an area 11′ 3″ wide. Because I knew the available lengths of 2 × 3s are 12′, 14′, and 16′, I set up the three sums underneath the drawing and had Robby compute the board residues we would have after cutting out our lengths. We selected the 14′ length as what we wanted because each board would yield a length and an end piece (we did not expect to need a 5′ high fence for our Scotch terrier).

In the process of explaining that sizing drawing and my objective for the project to Robby, I drew on the reverse of the 3 × 5 card with drawing A, the drawing B (see Fig. 3.2). The purpose was to locate our additional fence in a picture Robby would recognize (the fence behind the back of the house we lived in). As I gathered my tools and set up saw horses for working on, Robby drew his 'plan', drawing C. Contrasting drawing C with drawing B, you will notice it has vertical lacing similar to my earlier drawing (the hardware cloth, with its square mesh, was rolled up beside where Robby was sat drawing his plan), and his drawing is more detailed in showing the diagonal mesh of the chain link fence. Robby's drawing placed our planned construct in its setting, its context, but was useless for keeping track of the various dimensions one might want to use in calculations. I showed Robby, in drawing D, the kind of plan I had in mind and tried to indicate its purpose. I was surprised that his new idea of a planning drawing leaned so to verisimilitude.

Since Robby did not seem interested in my abstract drawing, we proceeded to the morning's rough carpentry. If you can't compute precisely the sizes of the parts you need, you can fall back on the handyman's technique of transferring dimensions, i.e. you lay a board next to the place it is to go and mark off the size it need be to fit. We proceeded in this simple, traditional way.

Transferring dimensions is frequently the procedure of choice in the repertoire of both the craftsman and the handyman; one reason is that the rustic carpenter is thus less vulnerable to the failings of his calculation skills; a second is that few things in this world, either natural or manufactured, are square; a third reason, decidedly relevant in this case of force fitting a wooden frame between a chain link fence and a masonry wall, is that you must sometimes worry about how your new manufacture relates to other things it is used with. Thus one lesson Robby could find in this project is that if you can't make an abstract plan, the concrete constraints may be enough, may even be your best bet, to finish the work at hand. Similarly in calculations, if you don't remember or can't understand some perfect algorithm, your commonsense knowledge helps you muddle through and may even be adequate to your needs. Though on the surface this may seem a negative lesson, it may be among the most important notions he can come to appreciate. Indeed, if particular solutions to problems are messy and inelegant, general solutions to really hard problems are usually impossible. As the following characterization of his learning shows, such a conclusion is quite congenial to Robby's own problem-solving practice.

THE CHARACTER OF NATURAL LEARNING

Flexible knowledge: extending a known solution†

At a parent–teachers meeting in the school both my children attend, the teachers were demonstrating the kinds of activities the children perform to develop their math skills. I managed to embroil myself in a minor conflict with a third grade teacher.

The woman had demonstrated to me the use of Dienes blocks as a concrete representation for children of the problems one encounters in addition (I think that is fine). I asked her if the children, using this medium, ever added from the higher order of columns of blocks to the lower. She noted both that they did and that she always instructed them to add 'from right to left'. 'Why?' Because, she explained even though they can add from left to right with Dienes blocks, they have to add from the right when they use paper.

Since she doubted that one could add left to right in the standard vertical form of addition, I exemplified the two-step procedure described in ADDVISOR. The teacher's counter-argument is shown by the sum she posed to me: 36 + 24 + 87 + 96; she claimed that one could not add it the same way. She noted that third grade children get to do such difficult sums by the end of the year. When I performed the sum in the same fashion, she argued that the procedure was too inefficient for computation. I let the matter lie, even

† From protocol 7, at age 7;3.

Drawing A

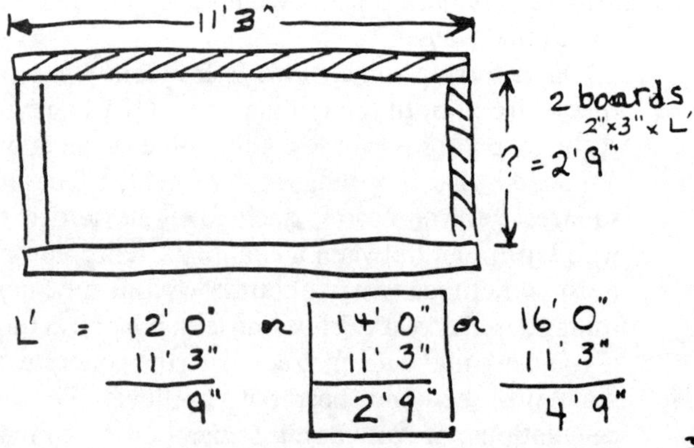

Drawing B

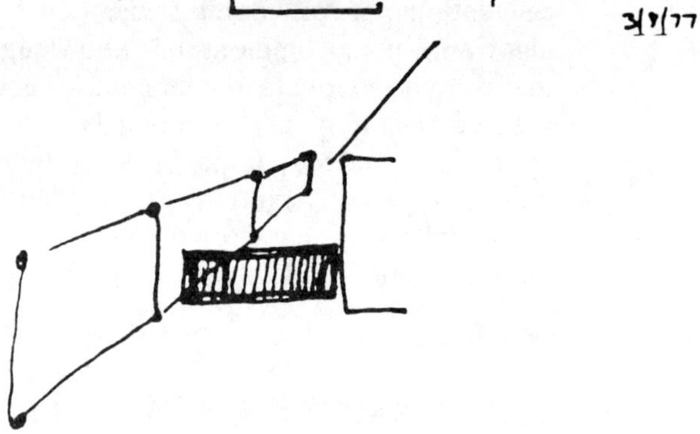

Drawing C

Drawing D

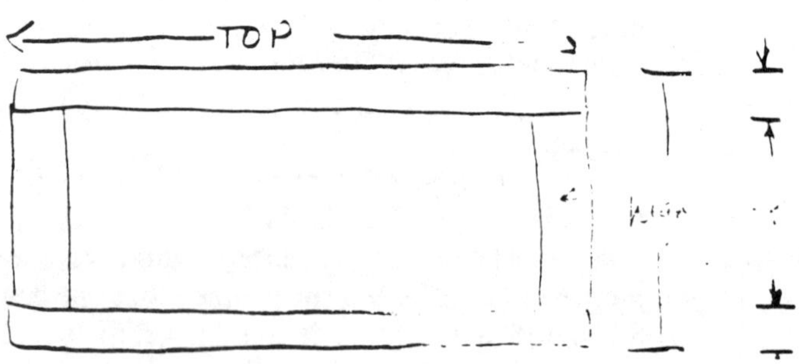

Fig. 3.2.

though I believe a short procedure of mysterious steps is not so good as a long procedure of well-understood steps.

The previous background material helps explain the particular puzzle I posed to Robby. I asked him to perform the sum over the four addends. Please note that I had never posed such a problem before and was unaware of his encountering any such (the most similar components he had confronted were cases where he had to add a carry with two other addends in a column). I explained to Rob that one of the teachers at the parent–teachers meeting (one he had never met) had showed me this problem as one she thought would be hard. I asked Robby to try it so he could develop an opinion on the claim.

Since Robby, after his experiences with ADDVISOR, had come to prefer adding right to left with carries, his exercise did not address directly the issue the teacher and I had discussed, but it did address how readily he could generalize the procedure he had learned for this new situation. As will be seen, it also show what sorts of number sense are useful guides in that process. I wrote the addends on a 3 × 5 card in vertical format with column division lines. Robby took the card and sat on a couch across the room.

Whenever he asked for help, I answered Robby's questions. He asked three times for verification of intermediate results. His first was 'Is 8 + 9 equal to 18?'† When told that the sum was seventeen, he asked if 17 plus 5 equals 22. The third query: 'Does 6 + 7 equal 13?' The order of the questions implies he was working on the sum from left to right. Next, Robby volunteered that 22 plus 23 equals 45 but that 45 could not be the sum because one of the addends ('the other numbers') is 96. He then asked a series of other questions about adding 'this and this and that and this'. I told him I couldn't see what he was referring to and that, if he wanted help, he should bring the work to me. He replied, 'No. I'm not finished'. After a pause, '243?', as he brought the work to show me.

I asked Rob directly if he had ever done any other problems such as this, and he said no. When I asked how he knew to treat the 2 of 23 like a carry, since he had never done it before, he said 'You don't have room for it here (in the units column); if you put it over there (in the tens column), your answer would be double big'. I take his response to mean that if the 2 were pushed into the tens column, the tens digits would be forced into the hundreds column; the answer would be 'double big' in the sense of having too many digits in the answer.

This little experiment produced a happy result. Such a constructive use of analogy is highly desirable and often effective. The child is in control; the personal extension of ideas he comprehends leads to the creation of more authentic knowledge. In this case, the minimal extension required to solve the problem may obscure the process involved. Next we turn to an example of calculation where analogy is used to bridge a wider gap.

† I believe this tentative sum is shown in Fig. 3.3 at the upper edge of the card to the right of the units column, as is a '5', which I take to be the intermediate sum of '2 + 3' from the tens column.

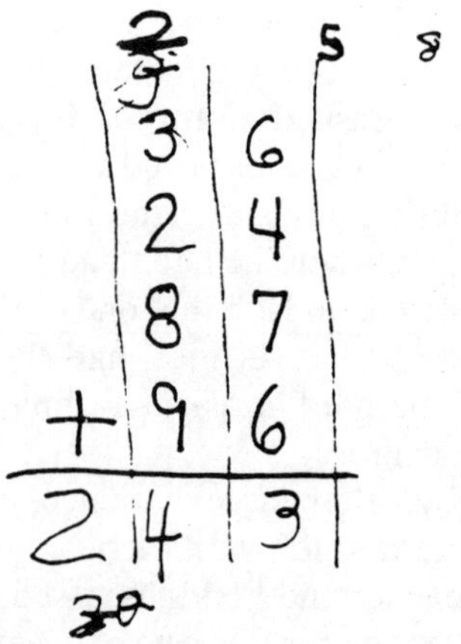

Fig. 3.3 — Robby's scratch addition.

Multiplication and money†

The term 'commutativity' as used in the following refers to only the most trivial sense — as a good trick that adults generally know can be applied to choose between two ways of performing a calculation. I try to 'tell' Rob that the reversibility of terms in sums, which he recognized, has its counterpart in multiplication. What Robby reveals of his methods of problem solving is more interesting than the prosecution of my agenda in the dialogue.

Bob: With adding it doesn't matter which way you do it.
Robby: Right.
Bob: What about with multiplying?
Robby: It doesn't matter.
Bob: Well, then, if we had a real hard multiplication like 25 times 4 . . .
Robby: That's 100.
Bob: How'd you figure that out?
Robby: 'Cause four quarters are a hundred.
Bob: What you did, Robby, was to turn these around, and you said, 'We've got 4 times 25'. That's a hundred because multiplication has the same property.
Robby: As money, sometimes . . .. Because four quarters of money is, uh. It goes like this: 25, 50, 75, 100.
Bob: You're right. Well, the point I'm getting at is that this commutative property, not mattering which way you do it, is true also of multiplication. Do you understand the property, Rob?
Robby: Yeah.

† From protocol 17, at age 7;7.

Bob: And that same property is true of multiplying?
Robby: Yes.
Bob: But is it true of subtraction?
Robby: No.
Bob: O.K. Thank you for working on this with me, Robby.

Two points deserve comment. First is the usefulness of capturing the agenda of the experimenter with mechanical recording in order to account for its impact on the subject. In this case, my focus on my agenda and lack of sensitivity to his behavior permits the argument that his deformation of a 'number' problem to a representation rooted in money-based experiences stands clear of my activity as an exampale of commonsense problem solving. The second point is the particularity of the computations Robby uses, as in his mapping of the twenty-five 4s multiplication onto his money-counting algorithm. While I remark that 'multiplication has this same property' referring to commutativity as in addition, Robby's understanding is that multiplication has the same property 'as money'. This example suggests that however much Robby's calculation abilities may appear as standard on the surface, his actual computations are a collection of clever tricks based upon his empirical understanding of number's application in a variety of domains. Commonsense problem solving could be much less a matter of inference than of trying to apply a variety of points of view — then following whichever one seems to work best. Abduction dominates induction. An important component of natural learning must be discovering in novel situation which components of already existing knowledge are useful, coupled with the construction of a solution based on this analogy. But if analogy is a central process in problem solving and learning, such processes must often go wrong. We now turn to examples of such natural confusions.

Natural confusions: place value and scale conversions†

The children had been exclaiming how they couldn't wait for Christmas. 'Tomorrow is Hanukkah. Maybe we should get one present for Hanukkah and all the rest for Christmas'. Believing this a strategem more than an upsurge of ecumenical spirit, I objected that we had no place for presents, had not even got a Christmas tree yet. Dragging my foot when they declared we should get one right away, I asked where could one go in this small apartment, what size should it be.

My reluctance was partly nostalgia. While the children ran about finding a yardstick, then a six-foot folding rule for measuring the least cluttered corner of our quarters, I regretted those impediments preventing me from returning home to harvest one of the hundreds of trees my wife and I planted years ago. Our trudging out in the near-Christmas snows to select, cut, and carry home our own holiday trees was a small but significant lost joy. The children quickly brought me back from this reverie.

† From protocol 11, at age 7;4.

Miriam ran to get a high stool while Robby stood with the six-foot rule against the wall and butted to the floor. Miriam proposed to measure from the top of the rule to the ceiling by climbing on the stool and using the yardstick. To prevent her from such an attempt, I suggested to Robby that he push the top of the rule to the ceiling and let Miriam measure to the floor from its lower end. Thus we arrived at our two measurements: 72 inches (read from the rule), and 24 inches (read off the yardstick).

Robby has been studying measuring at school, so he knew that we should be able to specify our maximum tree size in feet and inches. Miriam decided to leave this problem to Robby, who offered to work it out on my chalkboard. Writing down 72 and 24 in columnar format, Rob proceeded directly to the sum 96, which he noted was 96 inches. 'Now we have to do the feet'. I suggested we could figure out the feet an easy way. 'How many feet is 24 inches?' After a pause, Robby said 'Two'. and wrote down '2' on the chalkboard. 'How long is my rule?' 'Six feet', Rob replied; he wrote a 6 under the 2 and proceeded to the sum. Then he continued: 'We've done the feet. Now we have to do the inches'. Thus, adding in the 2 and 4 inches (from 72 and 24), he declared we should not gat a tree bigger than 8 feet 6 inches.

When I asked if he were sure his conclusion was correct, Rob informed me, 'Dad, I've had six weeks of measuring. I've done this a lot'. When I asked more specifically had he done *this* in school for six weeks, Robby responded, 'Well, I didn't count, but it was something like six weeks'. I continued to probe his calculation:

Bob: Rob, how come you added the 2 and 4 to get six inches?

Robby: The tens are the feet part [writing 'FEET' on the chalkboard over the tens column].

Bob: If the tens are the feet part, why don't 7 and 2 make 9 feet?

Robby: [After a pause] Then it would have to be 10 inches to the foot but that it's really 12 inches to the foot.

Bob: Well, if 12 inches is one foot and 24 inches is two feet, how much is two feet take away 24 inches?

Robby; Zero. Oh . . . you mean we used up the 4 in the 24 inches for the two feet?

When I agreed, Robby returned to the chalkboard, erased the 6 to 96 in his sum, and replaced it with a 2. Then, changing his mind, he complained:

Robby: Dad, you confused me [restoring the sum to 96 inches]. The tree has to be less than 8 feet 2 inches.

Bob; Where'd you get the 2 inches?

Robby: There, from the 72.

Bob: No. That's not correct.

Robby: It isn't? . . . I think we used up the 2 in the six feet, because it's 2, 4, 6, 8. Eight feet and zero inches!

Bob: How do you know you used up the 2 inches?

Robby: Feet are not tens, they are twelves.

We then, for verification, took out my folding rule and counted twelve inches for each foot. Robby kept track of the feet on his fingers and reached an even 6 feet as we came to the end of the rule.

As I jotted notes in my journal, Robby climbed into the chair with me and asked me to explain what I had written. So I did. He claimed that he never believed some of the things he had said, specifically that the 2 came from the 72. When I asked if it was OK to make mistakes, he said 'Yes' and laughed.

Using multiple representations†

After the specific confusion of feet and tens Robby had shown, I bought him his own six-foot folding rule, one with foot markings every twelve inches. Further, I decided to tackle his confusion directly by contrasting the addition of 76 and 29 with the addition of 7′6″ and 2′9″. We worked at my desk, which has a small chalkboard nearby. Even though we had the concrete support of various folding rules and a yardstick, Rob remained unable to develop any stable understanding of the relations and contrasts between these two number scales. Reviewing the transcript, I could accept the blame for this and admit that I tried to push too hard in asking him to add nembers expressed in yards as well as feet and inches. But the more important point coming from this didactic failure is that the ideas were not yet fully familiar, they were not yet fundamentally his own in some profound sense. Thus he could not use them to understand more novel problems.

Natural confusions: the multi-add algorithm‡

When one day Rob stayed home from school with a headache, he recovered enough with some aspirin and a long nap that in mid-morning he entered the study and asked me to join him at a frisbee game. I countered that inasmuch as he was missing school, I wanted him to do some thing 'academic'; I suggested he read the next chapter in Jane Goodall's book *The Wild Chimpanzees*. With that accomplished, somehow we fell into talking about multiplying. I told Robby if he had a good algorithm, one could multiply even very large numbers. Disbelieving that claim, and believing he knew how to multiply, Robby first set out a big number for me to multiply — then chose to tackle it himself. He used an algorithm of his own invention in which he multiplied digits within columns and carried tens digits from columnar products as one would in addition; I call this the 'multi-add' algorithm.§

Bob: 13 thousand, zero, three, zero, and you're going to multiply that by 13 thousand, 3 hundred and thirty . . . O.K. Now draw a line under that [See product 1 in Fig. 3.4]. What do you do now?

† From protocol 12; at age 7;4, a few days after protocol 11.

‡ From protocol 21, at age 7;9.

§ Miriam invented a similar algorithm years later, as reported in Lawler (1985, Ch. 7).

Robby: I do zero, nine, oh, nine, 2 [writing in the 'product', term by term, right to left, in product 2].

Bob: Zero times zero is zero. 3 times 3 is 9, 3 times zero is zero, 3 times 3 is nine, and 1 times 1 is 2? 'Cause that's what you wrote down, right? . . . What you're doing is putting a line down here, right? [as I draw lines between columns].

Robby: Yeah.

Bob: And you're multiplying every term. Your're saying 0 times 0 is 0. 3 times 3 is 9. 3 times 0 is zero. 3 threes is 9 and 1 times 1 is 2. O.K. Two things you can say about that. First: how much is 1 times 1?

Robby: 2.

Bob: No. That's 1 plus 1. O.K.?

Robby: Oh ho. 1 times 1 is 1.

Bob: O.K. So that's now correcter, right? Within the multiplication you did there, you did it fine.

Rob corrected the leading 2 of his product to a one. After I explained that I really didn't understand what he was doing, he erased the completed problem and began another.

Bob: Let me take a look at that. Holy smokes. 381 thousand, 156. Now do you want to multiply the same way on this problem as you did on the last one?

Robby: Yeah. Can I do it here?

Robby: This 6 twos — must be 12.

Bob: O.K.

Robby: Carry the 1.

Bob: [Probing] Why don't we just go ahead and do it column by column? Then worry about the carries after?

Robby: Daddy [exasperated].

Bob: Yeah.

Robby: Daddy [writing a 2 and carrying a 1]. 6 [the sum of the digit 5 and the carry 1] times 4 —

Bob: So you carry 1 and get 6 fours.

Robby: Yeah.

Bob: How much is that?

Robby: 6 fours . . . 12.

Bob: 6 fours are 12?

Robby: No . . . 18?

Bob: That's 6 threes.

Robby: [mumble] 22.

Bob: You said 22, but that's not correct, 6 fours are 24.

Robby: [writing]

Bob: So you put down the 4 and are carrying the 2? Hold on. Slow down, young fellow. You've got one and you're adding, you're carrying a 1.

Robby: Yeah. I should have carried the 2.

Bob: Let's take it slow and easy. Back up. So you'll cross that out [the 1] and carry the 2. What have you got now?

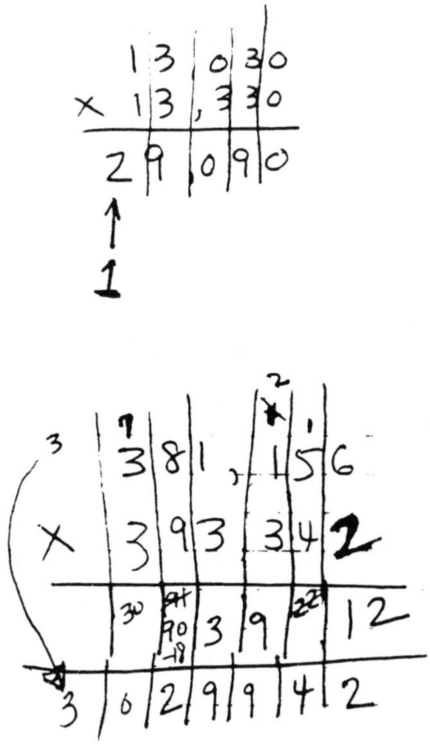

Fig. 3.4 — Rob's first multi-add problem (above). Rob's second multi-add problem (below).

Robby: [writing] 3 [1 plus 2 carried] times 3 is 9.
Bob: O.K.
Robby: 1 times 3 is 3.
Bob: Three.
Robby: [a groan for trouble encountered]
Bob: That's a good one, isn't it.
Robby: 8 times 9 is . . . 8 nines . . ..
Bob: Is that too tough?
Robby: Yeah.
Bob: Try 10 nines and take away 2 nines.
Robby: 10 nines. What's that? 91?
Bob: 10 nines is what?
Robby: 91?
Bob: No.
Robby: What . . . 90?
Bob: If it's 9 tens, it's 90. Take away 2 nines, which is 18.
Robby: 18.
Bob: So it's 90 take away 18.

Robby: 62?
Bob: That's close. It's 72.
Robby: Aggh.
Bob: So you put down the 2 and carry the 7? [describing Robby's action, not directing it].
Robby: 7 and 3 is 10. 10 threes.
Bob: Are what? . . . Zero, thirty. So you put down the zero and carry the three . . .. What you've done is to invent a new kind of way of doing this. What you're doing is very interesting but it's not multiplication the way other people do it.
Robby: I do it this way . . . It's easier.
Bob: I'm sure it is, though I don't know that it's so very easy.

As our problem solving continued, I tried to explain to Rob why his algorithm, even if he should perform it correctly, would be different from what other people meant by multiplication. The attempt met with no good success. His own experience convinced him that multiplication is like addition — didn't he multiply two numbers by repetitively adding the first the number of times represented by the second? Given that opinion, should not the addition algorithm for large sums be modifiable for large multiplications?

The analogical extension Rob imagined was primarily a syntactic modification based on a correct, semantic comprehension of the relation of the parts. If commitment to his understanding was a weakness in this case, the weakness was very strong. I conclude that he needed a different sort of experience, one where the elements and their relations could become as familiar to him as were those of addition but through which he would be able to see that a multiplication algorithm must be quite different from his addition algorithm. That is, he needed some alternative model from which by a constructive use of analogy he could create for himself the standard multiplication algorithm.

INSTRUCTION AS INVENTION

Memorization and engaging games†

For some time, Robby has been asking me for help in memorizing single-digit sums. Whenever I have to memorize relations which demand an unthinking but accurate response on my part, I use flash cards and drill to embed the process in my mind. I don't want to emphasize this method with Robby now because I want to avoid his developing the idea that this is what 'learning' is. These values confronted me with a problem: how can I engage the children in an activity wherein addition of single-digit sums is an integral element and wherein the demand for that addition is repeated in rapid cycles.

One game I know with these attributes is 'shooting craps'. This is a

† From protocol 9, at age 7;3.

gambling game played with two dice. The players put money in a pot, and one person rolls the dice. Should he roll a sum of 7 or 11, he wins the money in the pot. Should be roll some other sum, that becomes his 'number,' which he attempts to match in subsequent tosses. In these succeeding tosses, should he roll a 7 or 11 before he matches his number he loses and the turn passes to another of the shooters. After introducing the children to the game, I found they were fascinated (they have played consistently for several days), and they even introduced the game to visiting friends.

The only drawback I've found in the game (aside from its centuries old ill-repute as the pastime of gamblers and ne'er-do-wells) is its limited range of sums, i.e. with six faces on each cube and both numbered 1 through 6 inclusive, the highest sum is twelve. Robby has mastered those sums adequately but requested help with sums such as 6 + 9 and 7 + 8. The regular pentagon generates a regular solid of 12 faces, and two of these could yield sums to 24. (But such would be hard to make, would roll too well, and would provide a less sharp definition of the dot patterns than does the cube.) I preferred the simpler alternative of modifying the patterns on a set of standard dice. I now have two dice with these patterns of dots (4 through 9 dots):

4 5 6 7 8 9

The possible ranges of sums from mixing standard and super-dice are 2 through 12 (two standard dice), 5 through 15 (one standard die and one superdie), and 8 through 18 (two superdice).

The children have been playing 'super-craps' today. From the playroom I hear Robby lament 'Oh no, seventeen' (with a different range of sums we picked 13 and 17 as numbers analogous to 7 and 11) and from Miriam, '15's my number'. Using the higher range of dice has slowed their game marginally but has not made it unattractive.

A final concern for such a technique of helping children learn single-digit sums: what sorts of representations are involved in their addition processes in this task? What I see is the children translating the problem into their standard addition procedures. That is, Miriam starts at the larger number and counts up to the sum. Robby apparently estimates and refines, e.g., '9 + 6 is 16, 15'. This I take to be counting back from a sum, of the smaller digit and the larger rounded to 10, by the difference of that larger digit and 10.

A week after the introduction of my children to 'super-craps' their mother and I met Robby's teacher for the first of periodic conferences. We discussed the problems he had with other children in the class (those typical for a new boy moving into the neighborhood). The teacher volunteered that she had been worried about Robby's inadequate command of 'number facts' until the last test, wherein he responded accurately and rapidly to single-digit addition problems. The test was given five days after the children had been introduced the shooting craps.

A subsequent variation on the use of the dice was in subtraction problems. The green superdie pattern ranged from four to nine. I made a mixed set of one green die and one standard red die. The problem posed was to take away the count of dots on the red die from the dots on the green superdie. Since the problem was not implicated in any game, it was engaging only for the challenge it offered. Robby has played with the dice occasionally and performed the subtractions. Once in a while, the subtractions produced a result less than zero (for example, a red 6 and a green 4 resolve to minus 2). When he expressed those negative differences as 'zero minus 1' and 'zero minus 2', I asked Robby if that was the way they discussed such numbers at school. He replied that they hadn't talked about such numbers at school yet.

Comparing this game with the earlier computer experiences of ZOOM and ADDVISOR, I judge playing with the superdice as more like using ZOOM than ADDVISOR. ZOOM provided a bounded world where drawing experiences required developing familiarity with a numeric scale for eventual satisfaction of the child's objectives. ADDVISOR has a different quality, that of an articulate explication of a representation, one highlighting the linear independence of columnar sums and their permitted interactions. Its limitations derived from the sharp focus on a particular representation, which was also the basis of its usefulness. We turn now to another program with the same character.

Guiding constructive analogy†

For the past six weeks or so, Robby has been asking me to help him learn multiplication. He needed a richer conception than 'it's the same as adding the number so many times'. When I asked him why he wanted to learn to multiply, Robby explained that he didn't want to have to learn it in school; pushed further, ne noted some multiplication problems were within the next few pages of his math book. The first question is one of representation. What is the prototypical problem in which multiplication is used? It is most usefully exemplified as the process by which one can find out how big is some rectangular area. Such a description is coherent with the uses of the Cuisenaire rods and the Dienes blocks Robby has available in school.

Robby's explanation of what Dienes blocks are and of what numbers he can multiply are shown in the first two items of Fig. 3.5. When asked if he had been taught these specific products in school, he said, counting on his fingers, 'It's 10, 20, . . ., 100; and it's 20, 40, 60, 80, 100!' Beyond this compatibility, I saw Robby's familiarity with the 'flat' (the 10 × 10 square of the Dienes block set) as a means of helping him develop first, an estimate of how big a product is and second, a tool for computing how big a product is.

As a consequence of Robby's interest, I wrote a program for the Logo computers to help him understand what multiplication is. The intention of this program is that it be a tool, not so much for computation as for flexibly representing a computation. The program is invoked by typing its name and

† From protocol 13, at age 7;4.

Fig. 3.5 — Representing Dienes blocks (above). Spontaneously calculated products (below).

two numbers, e.g. TIMES [12 13]. The resulting pictures of its execution are shown in Fig. 3.6. TIMES displays its multiplicands as two perpendicular rows of tiny squares. The product is represented by extending parallels to each row of squares to complete a rectangle. Next, at the child's direction, the program breaks up the rectangle into a grid of 10 × 10 squares and 'left overs' to help the child estimate the product.† A first order estimate of the product is made by summing the 10 × 10 squares. The product can be computed by adding together the four single-digit products. That is, for AB × CD, in Fig. 3.6, the product is: (A × C × 100) + (B-× D) + (A × D × 10) + B × C × 10). In a final step, TIMES displays these four partial products at the bottom of the display in rows, each of which may contain as many as one hundred unit blocks.

When Robby first encountered TIMES, my purpose was to see if this sort of tool could help him grasp a representation which would greatly expand his conceptual range. I have come to suspect, since my work with ADDVISOR, that such an achievement is inspired by an immediate exposure of the child to the power of the ideas involved. Let me be more concrete. We usually

† This representation was suggested to me by Greg Gargarian.

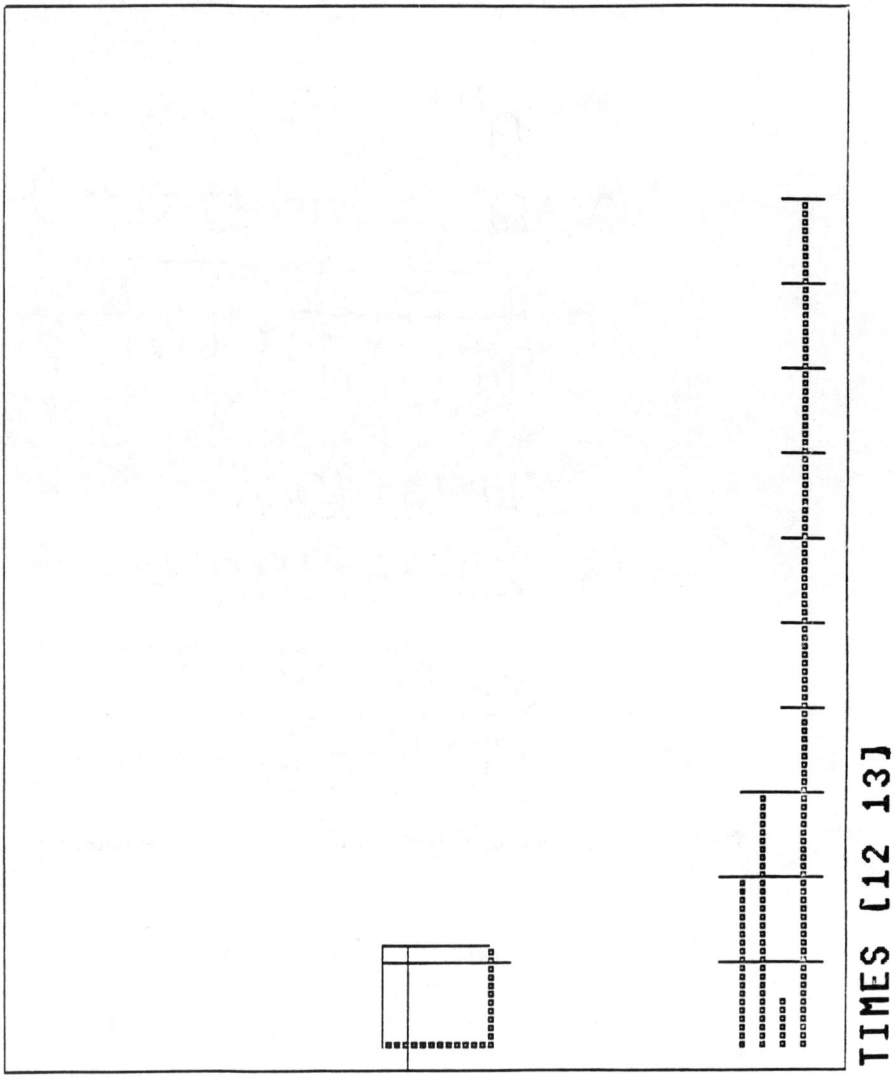

Fig. 3.6

teach children to multiply numbers such as 2 × 3 and 6 × 2 because 18 × 21, as computed by the standard algorithm, is difficult to keep track of and impossible to explain. Thus, in:

$$\begin{array}{r} 18 \\ \times\, 21 \\ \hline 18 \\ 36 \\ \hline 378 \end{array}$$

How does one explain that there are two (and only two) intermediate products and that the lower one is shifted one place to the left? Numbers of this magnitude, however, are such as Robby freely choose to trying multiplying.

Bob: What numbers for TIMES?
Robby: 20 times 20.
Bob: This program tries to show that a good way to think about multiplying is like trying to figure out how big rectangles are . . .. You can think of one number you multiply with as being one side and the other number as being the other side. And the product, the product is like how big the whole area is in here. That's what multiplication is mostly used for, to do things like that. Does that make any sense?

Robby responded with sullen looks and said he was reluctant to use the program. He did agree, however, to use it *one* time. After I invoked TIMES with the numbers 20 and 20:

Program: [Displays the question, 'How big is the product?']
Robby: Do I have to answer that question?
Bob: Can you make a guess? . . . You don't want to guess?
Robby: No.
Bob: Don't. The next thing, when you press the space bar —
Robby: [presses the space bar twice]
Program: [Displays the product as a rectangular array. See Fig. 3.6]
Bob: This is to help us guess. It makes big blocks that are part of the answer.
Robby: Oh. 400 could fit in there [smiling].
Program: [Continues on to display the product in stacks of length 100.]
Bob: Right. That's the answer. Each of those blocks, like a hundred, will fit in there.
Robby: Oh, The ans — 20 times 20 is 400!
Bob: Yeah. You can see 400 will fit in there 'cause each one of these [gesturing towards screen] is like one of the flats of the Dienes blocks. And this, at the bottom, is just another way of lining up 400 things — with a hundred in each layer of the pile . . .. Want to try another one?
Robby: [smiling] O.K. . . . [keying TIMES18 21].
Bob: 18 by 21. Wow! That's going to be a tough one.

Robby continued working on this problem. He appeared to use the rectangular array of blocks to develop an estimate of what the product should be then use the linearly stacked blocks as a basis for calculating the product. For example, 18 times 21 can be calculated as 200 plus 160 plus 10 and 8.

This program focussed on providing a child with a representation for a problem which could be meaningful in terms of his past experience and could be useful in the sense of bringing within his grasp problems otherwise beyond his reach. Like ADDVISOR, TIMES is focussed on the articulate explication of a representation through its use in solving particular problems. What such programs gain in specificity — as contrasted with ZOOM, for example — they lose in flexibility.

CONCLUDING COMMENTS

If we ask what we can gain as scientists from this exploration of a child's natural learning, I believe we can come away with a sharpened perception of how difficult it may be to answer the core developmental question, 'What are the precursors of the mature forms of commonsense knowledge?' In addition, I am convinced that it is essential to capture the richness and centrality to learning of the social milieu. Concerning education research, attention to powerful problem-solving ideas, such as the assumption of linearity, and to multiple representations and their interactions will be valuable. Moreover, we can now try to instruct by inventing new little worlds of experience which relate directly to the particular experiences of individual children. This use of computers as a flexible medium for creating learner-specific, crisply articulated experiences is likely to be the most important contribution of information technology to education — if instructors become masters of the programming languages they will need to use.

With respect to methods for further study, the clearest need is to go beyond anecdote. This argues the profit of more extensive data collection, even of going so far as to capture a complete record of a subject's behavior and learning for specific domains for an extended period of time. Even though the instructor may be an engaged experimenter, mechanical recording of his behavior can help the analysis to account for the effect of his behavior on the subject of the study.

Such a study has been undertaken, in fact. In the period following the collection of these protocol materials, I completed an extensive and intensive study of the learning of my two older children. One report of that work is the analysis of my daughter's learning in *Computer Experience and Cognitive Development* (Lawler, 1985). The concluding chapter of this part, 'Extending a powerful idea', represents a developed analysis of a part of Robby's activities during the same period of time.

4

Extending a powerful idea†

The argument

Mathematics is much more than the manipulation of numbers. At its best, it involves simple, clear examples of thought so apt to the world we live in that those examples provide guidance for our thinking about problems we meet subsequently. We call such examples, capable of heuristic use, *powerful ideas*. This chapter documents a child's introduction to a specific powerful idea in a computer environment. We trace his extensions of that idea to other problem areas, the first similar to his initial experience and a second more remote from it.

Introduction

The new availability of computer power to children in schools poses forcefully the question of the computer's role in education. Here we present a case study of computer-based learning that goes beyond drill-and-practice and game playing to show how particular experiences carried a specific idea of general applicability into a child's mind and how this idea was effective subsequently in his freely chosen and self-directed problem solving. The stepping of variable — by which we mean the development of and the decision to apply a systematic mental procedure for isolating and incrementally changing one of several variables — is an idea of general applicability. Genevan psychologists have noted this idea as a very important one in the

† An earlier version of this chapter appeared in *The Journal of Mathematical Behavior,* Vol. 3, No. 2 (1982).

configuration which leads to the systematic thought of the adolescent (Inhelder and Piaget, 1958). the stepping of variables is the idea whose history we will trace in one child's mind.

The subject of this case was Robby, my son, just turning eight years old at the time of the study. With his sister, Miriam, two years younger, he participated in an intensive six-month study at the Logo project of the MIT Artifical Intelligence Laboratory (Lawler, 1979 and 1985). Robby had visted the lab many times in the preceding years and had frequently participated in earlier studies. He brought to this work two mental predispositions which are relevant to what follows. First, he was inclined to call upon symmetry as a generative idea (this observation will be clarified subsequently). Secondly, his approach to problems was surprisingly systematic for one his age.† Before we turn to the case material wherein systematicity plays its role, we must first present some folk-history on the procedures Robby encountered.

POLYSPIRALS AND VARIABLES

Imagine that one day a child invented the "squiral" while trying to draw a square maze. The Logo turtle moves forward, turns a right angle, then repeats these actions, increasing the distance of the forward move with each repetition. In getting the angle 'wrong' while attempting a square maze, the child discovered that if the turtle turns through an angle near but not equal to 90 degrees a four-armed spiral emerges from the drawn shape (see the maze and squiral in Fig. 4.1).

Emergent effects such as this appear regularly in turtle geometry, and are very striking. The nature of the Logo language, and the spirit in which students use Logo, make it easy and natural for students to change programs in various ways. The 'square maze' and 'squiral' programs were modified, by students, to produce the "polyspi" (short for polygonal spiral). The actual Logo program that produces these figures is as follows:‡

```
TO POLYSPI
FORWARD :distance
RIGHT :angle
MAKE "distance (:distance + :delta)
POLYSPI
END
```

The general meaning of these Logo commands can be inferred fro the example. Details are not important for the purposes of the present chapter,

† For example, at age 7;8;8 on Piaget's 'bead families' task (Piaget and Inhelder, 1975), after attempting to arrange combinations of five things taken two at a time by an empirical procedure, Robby spontaneously started the task a second time, grouping his bead couples in five groups by the color of one bead and joining with each base-color bead another one of a different color. Such systematicity is not usually met until the age of ten or twelve years.

‡ We signify references to variables by preceding them with quotes, executed procedures are referenced in capitals, and definitions of terms are enclosed in double quotes ":" is an operation in Logo which means 'give me the current value stored for this variable name'.

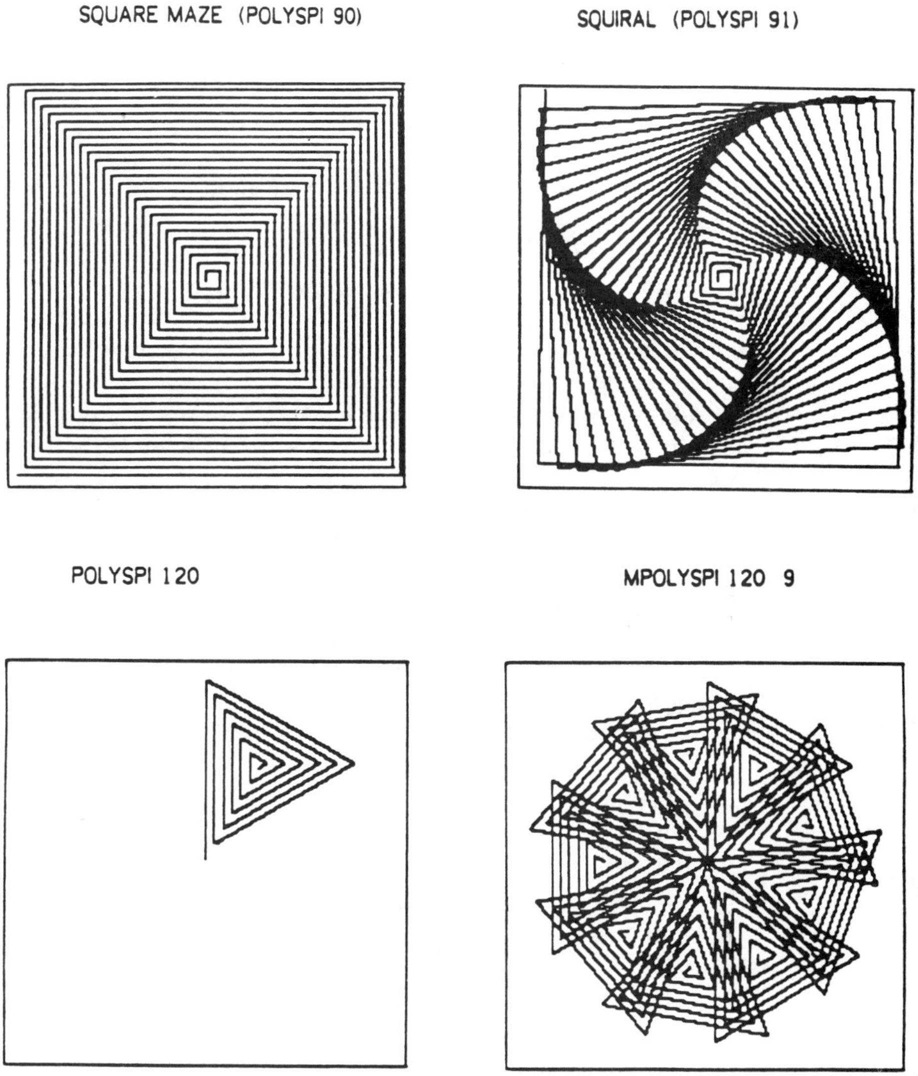

Fig. 4.1 — Basic mazes and squirals.

except to note that "distance, "angle, and "delta are variables, and must be initialized. This means that the computer must somehow be told what number to associate with the named variables "distance, "angle, and "delta.

The fifth line in the POLYSPI listing is of special importance. The procedure POLYSPI *calls on itself!* A feature of the Logo language makes this possible. Notice, however, that before calling on itself, the procedure has *changed* the value of one of the variables; this is done in the line:

```
MAKE "distance (:distance + :delta)
```

Consequently, the result of POLYSPI calling itself is to execute the procedure another time with the value of the distance variable incremented; the procedure stops when the turtle goes off the edge of the display screen. A number of the designs produced by the POLYSPI procedure from the near-regular polygonal angles are very pretty. The emergent effect of such designs can be compounded, as by a procedure I composed, MPOLYSPI (short for multiple polyspi), to make even more complex and attractive designs (see the triangular POLYSPI and its nine-fold compounding in Fig. 4.1). Emergent effects in turtle geometry cover a range which permits mutual engagement and learning by both children and adults — and thus they became a topic of exploration in many of the research sessions Robby and I spent at the Logo lab.

Robby's introduction to variables went forward in several small steps. Initially we played games with a set of labelled matchboxes and card-written commands for changing or examining the contents of the boxes. We presented the image that global variables were functionally like little boxes with contents that one could examine and change by commands of the Logo language. Consequently, Robby's first use of polyspi procedure involved a "set-up" procedure whose functions were to clear the display screen of the previous design and to permit his keying of initial values for "distance, "angle, and "delta. Subsequently, we explained "input variables" as a keying convenience which permit an integrated set-up procedure. The form of the "polyspi with input variables" is this:

```
TO POLYSPI :distance :angle :delta
FORWARD :distance
RIGHT :angle
POLYSPI (:distance + :delta) :angle :delta
END
```

(Here the incrementing of distance is implicit in the polyspi self-invocation.)

A primary intellectual challenge in exploring polyspi designs is how to impose some comprehensible order on their considerable variety. Notice, now, that the polyspi procedure with a zero value for delta will draw polygons. For such polygons drawn with angle values dividing evenly into 360, the polygons are regular and closed; after turning through 360 degrees, the turtle retraces its original path. In second grade, Robby had learned the names of some geometric figures and the number of their sides. For example, he knew that a regular hexagon has six sides. When the input angle is that of some closed polygon, e.g. 60, the polyspi procedure with small positive values for delta draws figures we called "mazes". We distinguished between such mazes and other shapes. Robby and I together produced a "family of mazes", i.e. a collection of the regular polygonal spirals with three through eight sides. We printed these designs and displayed them on the wall. My intention was that individual members of this series of figures could serve as "anchors" for further exploration, both connecting to

Robby's previous knowledge of geometric forms and serving as reference bases from which other shapes could be seen as variations.

In the next Logo session on this theme I presented explicitly the objective of developing "families of shapes" and showed Robby several examples of such shape families. The first, reproduced in Fig. 4.2, shwos six

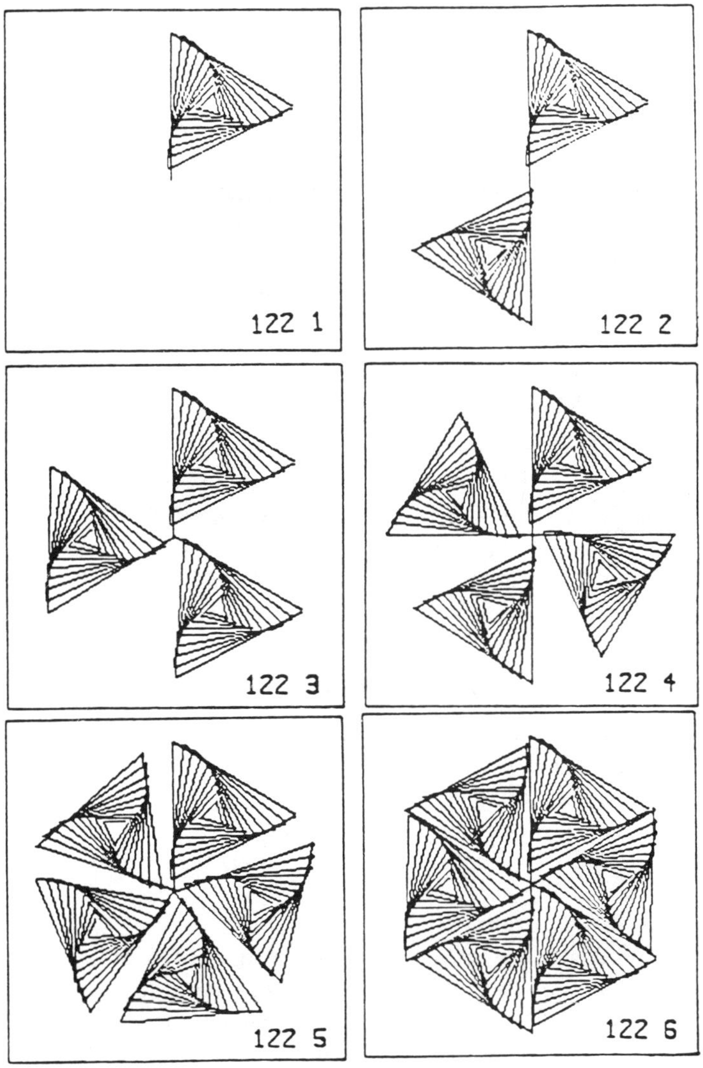

Fig. 4.2 — Multiple polyspirals: a shape family.

shapes made by incremental change of the MPOLYSPI 122 "folding factor" from one to six. Thus the single sub-figure, a POLYSPI 122, is repeated an additional time in each of the five successive designs. Similarly, a second example of a shape family (not shown) displayed changes in the six-fold MPOLYSPI 122 as the value of delta was reduced from seven to one. The 'lesson" I professed with these examples was that focussing on the systematic changes of a single variable was a fecund method for understanding the results from the complex interaction of several variables. Whether or not Robby accepted my "lesson" or used what I showed him in another way is an issue we will discuss subsequently.

Robby had the opportunity to construct his own shape family. (I had been careful to leave him the most interesting variable for his changing). I proposed building a shape family around one of the mazes as an anchoring value. Robby selected the hexagonal anchor and enthusiastically created and printed the designs of Fig. 4.3, then hung them in order above his desk. In subsequent sessions, he constructed a shape family anchored at 90 degrees, first increasing the angle value and later (beginning at 85 degrees) approaching the anchor from below. This concluded the didactic phase of Robby's introduction to shape families.

A POLYSPIRAL VARIATION

The increment input to the POLYSPI procedure (the variable called delta) is applied to the first of the two other variables, "distance and "angle. What happens if delta is applied to augment angle instead of distance? Few people have any intuitive answer for such a question and most become easily confused in trying to imagine what design would be created. With small values of dalta (and with an initial angle of zero), the turtle will move off in one direction and gradually spiral into a node. (For this reason, the procedure was named "inspi" by its originator, Marvin Minsky):

```
TO INSPI :distance :angle :delta
FORWARD :distance
RIGHT :angle
INSPI :distance (:angle + :dalta) :delta
END
```

(see several examples of basic Inspi designs in Fig. 4.4.)

When he was first introduced to inspi designs, Robby's inclination was to vary "angle. After trying INSPI (10 0 2) at my direction, he executed INSPI (10 90 2). (The effect of these input values, shown in Fig. 4.4, is to alter the center location and orientation of the design.) When I tried to discuss changing other variables, Robby's personal agenda came forward in comments such as these: 'Hold it, Dad, all I want to make it do is go the other way'; 'I want it in the exact same direction, but opposite'. After a few false starts, I followed his lead:

Robby: How do you get it to go the other way?... It goes right. I want to make it go left.

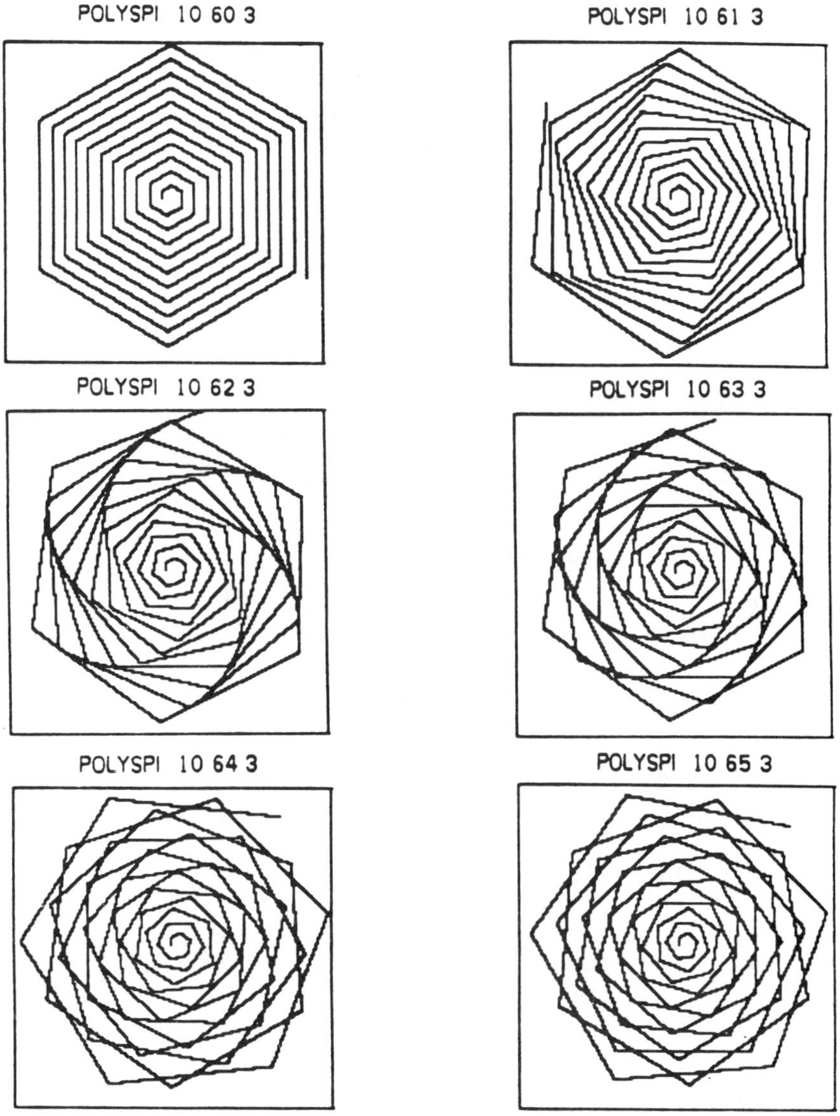

Fig. 4.3 — A hexagon-based shape family.

Bob: We'd have to change the A procedure [we had renamed inspi A because it changed angle values].

Robby: Why don't you make a B procedure, to make it go the other way?

Bob: Why don't you?

Robby: [a complaint] I don't know how!

Bob: You copy the A procedure, but where it says 'right' you change it to 'left'.

After creating the symmetrical procedure, Robby still inclined to vary

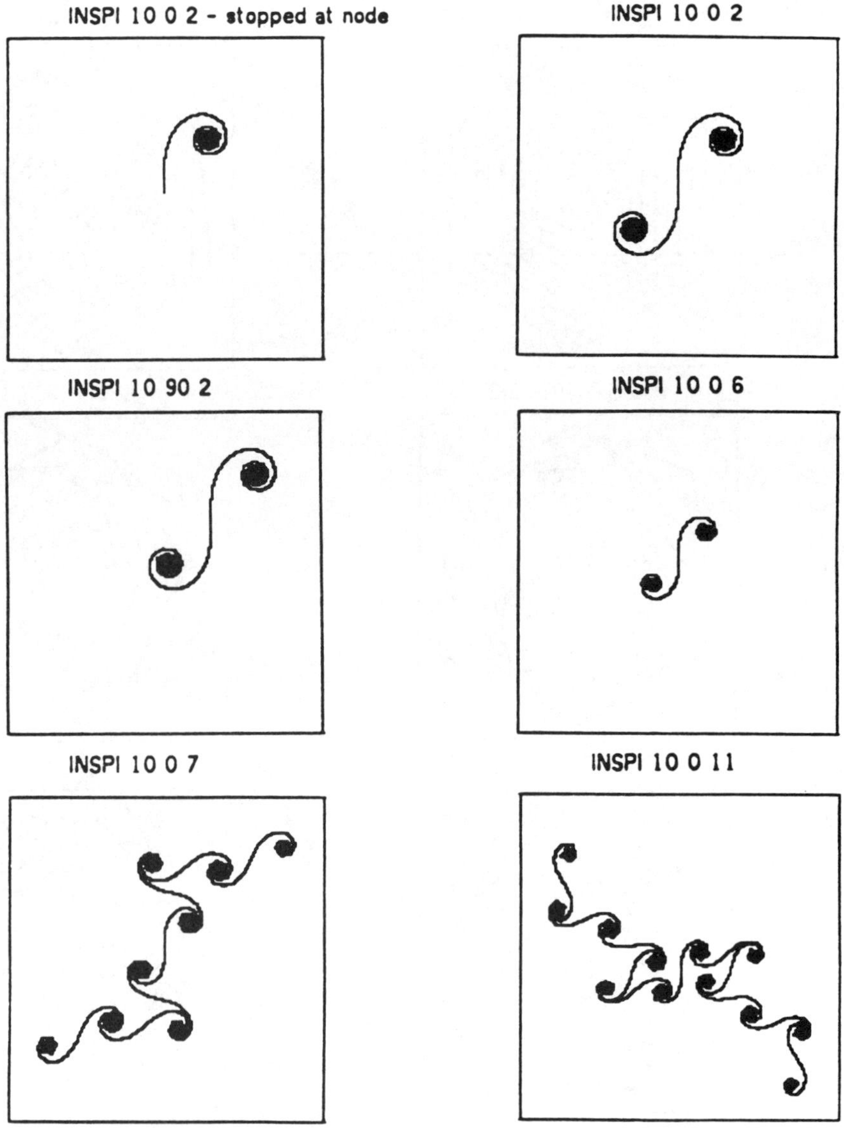

Fig. 4.4 — Six drawings made by the INSPI procedure.

"angle. I intervened to focus his attention on delta as a possibly potent variable, but he could not be interested at all. When he refused to follow my forecful suggestion to exeute INSPI (10 0 7), I did it myself with those inputs and had the satisfaction of hearing him admit, 'I wish I had done that'.

Robby now began to make the symmetrical inspi designs of Fig. 4.5, using his B inspi procedure with inspi procedure A. As the complex symmetries of the two-fold INSPI (10 0 7) developed, he exclaimed, 'Isn't that wild'! I offered 11 as the next delta candidate. After printing out the

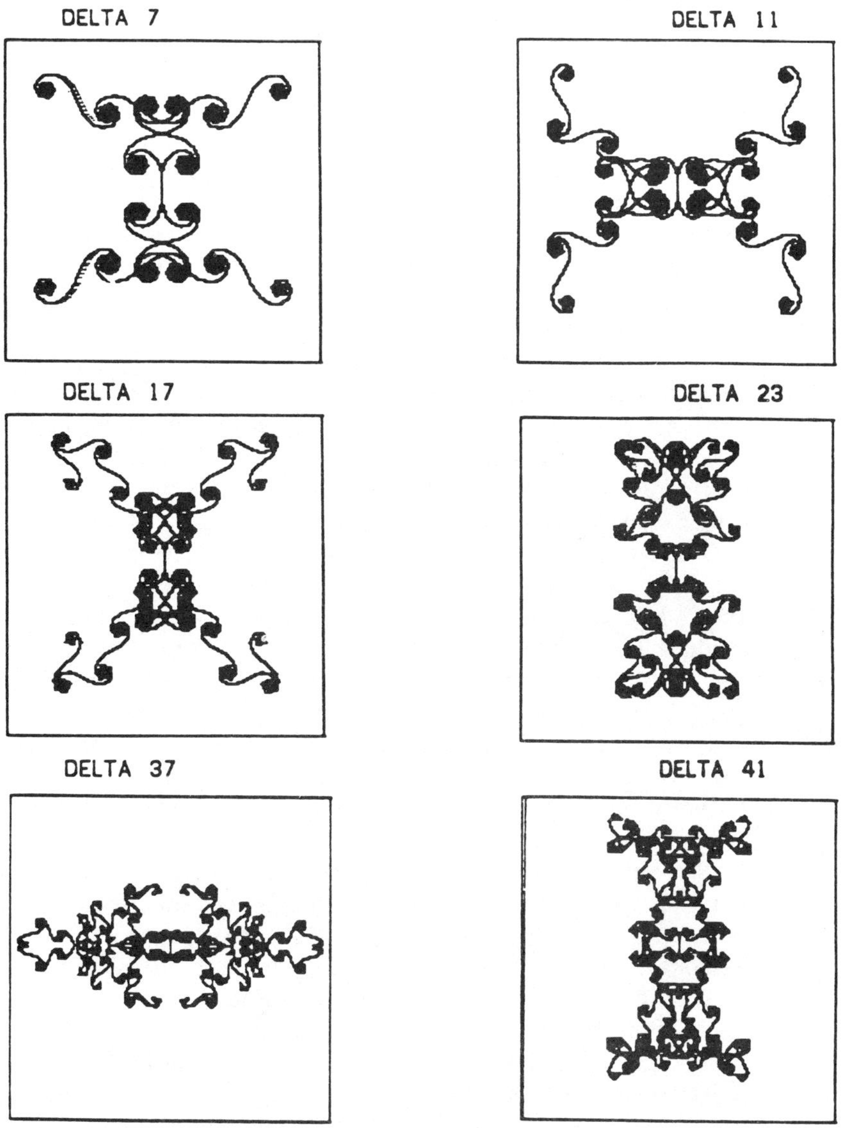

Fig. 4.5 — A selection of Robby's symmetrical INSPI designs.

second symmetrical design, I proposed (would that I had bit my tongue) this speculation:

Bob: Why don't you try the next prime number? It turns out — and I never would have guessed it — that prime numbers —

Robby: [interrupting] Do this? [referring to the symmetrical INSPI (10 0 11)]

Bob: Why don't you give 13 a try?... I'm not quite sure, but that's my speculation, anyway. I'm sorry I told you that.

Mayb I should have let you have the chance to figure that out.
Robby: Yeah... but I don't know the prime numbers very well.

After executing the symmetrical INSPI 13s, Robby concluded, 'Every time we do it, they're getting super-er'. We conclude that Robby's strong confidence in the heuristics of pursuing symmetries in design was amply confirmed. This use of a heuristic is one clearly focussed on the objective of generating interesting designs. If there were any motive to understand better these inspi designs through symmetry it must have been relatively insignificant compared to his delight in creating them.

My uncertain speculation that primes had something to do with creating interesting designs inspired Robby to produce a complete set of symmetrical inspi designs for all the primes betwen seven and fifty.† One might imagine he did so because he is a suggestible boy and easily led. The opposite is more nearly true. Two examples stand out. At my direction, he tried some large-value deltas. When I called his attention to the puzzle of the minimal design (a small straight line) made with the initial angle zero and delta 180, Robby continued from there with elegant symmetrical designs, based on values of 187 and 206, then decided, 'I think I'll go back to using primes'. After producing designs for delta values of twenty-nine and thirty-one, he decided that thirty-three was not a prime and rejected it as a candidate value.

Knowing that thirty-three would produce an interesting design, I pushed him to try it, but he refused. 'No, I'm going to use only primes'. He then figured out the values of primes thirty-seven, forty-one, forty-three, and forty-seven and completed his own creation — the family of symmetrical, prime-based inspi designs.

What do we make of this material? Most obvious is that the world of experience confirmed the value of Robby's heuristic, 'try symmetry', as a generator of pretty designs. Second is that even though the inspi procedures led to results less intuitively accessible than those produced by the polyspi procedures, Robby explored this world of inspi designs in a systematic way that amounted to a first extension of the idea of variable stepping. That is, Robby settled on delta as the most potent variable for these designs and followed my 'prime hypothesis' to generate the next candidate delta value for creating a design. I believe we can infer that his experience of this inspi world confirmed the value of stepping variables as a heuristic.

The symmetry heuristic is good for generating designs; what is stepping

† The apparent complexity of an inspi design is determined by the sequence of values of the angle variable. Most significant is the remainder left when the increment value is divided into 180 and 360. As the simplest example, consider the case where the initial value of "angle is zero and "delta is ten. After some iterations, the turtle will turn right through this sequence of degrees [170 180 190]. RIGHT 180 turns the turtle around completely. RIGHT 190 is equivalent to LEFT 170. Thus, at the first node, the turtle begins executing steps which invert, in reverse order, each preceding step. Since ten divides into 360 with no remainder, we can see that there will be another node at an "angle value of (180 + 360) degrees, after which the turtle will once more retrace its path. The primes are merely a subset of the numbers which don't cause the turtle to retrace its path after the second node.

variables good for? Did Robby appreciate this as a second heuristic for generating pretty pictures or did he see it as a way of organizing the world to understand it better? This distinction is one that *we* might call upon in judging a possible claim that Robby learned a heuristic that is good for some specific purpose. I believe rather that *Robby learned a heuristic that was specific with respect to activity but vague with respect to purpose.* Although he may have begun to apply the heuristic to generate interesting results, his ability to select and order them through the 'prime hypothesis' helped define what was interesting about them. *The final outcome for him was the better comprehension of something worthwhile understanding (in terms of his judgment of what was worthwhile).* This point is illustrated in the final incident reported here.

BEYOND THE LABORATORY

Because Robby lives with me, it has been my privilege to observe how these experiences at Logo were reflected in his later problem solving. Some six months after our study at the Logo project, a parent visiting his third grade class introduced to Robby the 'paper-rings puzzle'. This bit of topological magic leads to the 'squaring' of two circles. (It is a puzzle in the sense of a surprising result.) I recommend you try it. Here's how it's done:

(1) Cut two paper strips of equal length (eight inches will do).
(2) Draw a line down the middle of each.
(3) Bend each strip of paper into a circle and tape the juncture.
(4) Join the circles at tangents perpendicular and tape the juncture.
(5) Cut around the middle line drawn on each circle.

When two strips of equal length are so connected and cut, the surprising result is that the strip-halves end up taped together as a square.

Robby enjoyed this activity when shown it. Several days later, I removed a pattern of strips he had made from a paper on my clip-board. When I interrupted his reading to give him the sheet of paper, Robby recalled the game and quietly took it up on his own. He was very happy when the procecure produced a square and showed it to his mother and me. We neither paid much attention. Going on to three circles, Robby cut two of the three along their mid-lines. He judged (in error) that he had finished by finding a square with a bar (a double-width strip) across the center. It lay flat. Still no one paid attention. He went on to four circles. When he cut all the mid-lines, what he got was a confusion of flopping paper strips. I advised him to try getting it to lie flat. He was delighted when he achieved this goal and subsequently taped the paper stips to a large piece of cardborad. The resulting shape is shown in Fig. 4.6.

But why stop at four? Robby went on to cut and connect five circles. When cut, the five circles separated into two identical non-planar shapes. He taped these to another piece of cardboard. On trying to tape the floppy

Fig. 4.6.

figures made from cutting six rings, Robby succeeded with great effort. He decided the problem was getting too complicated to be fun and quit.

When I recalled his attention to the figure made by cutting two of the three circles and pointed out that the middle bar of his figure was double-thick, Robby agreed he had cut only two circles. He was immediately that this square would divide into two rectangles. 'The five's made two too. Hey! I've got a new theory: the odd-numbered circles make two and the evens all stay together.' Robby could not prove his conjecture, but in the course of one discussion when I asked how he had gotten the idea of this exploration, he explained, 'Dad, it's just like what we did at Logo with the shape families. I changed one thing, a little at a time'.

Robby's explanation witnesses that he conceived of his exploration in terms of the past Logo experience. This does not imply that the Logo shape families marked his first encounter with or use of the idea of controlled changing of a single variable. We may infer, nonetheless, that he owned an example of this idea crisply aplied to a complex but comprehensible range of interesting phenomena, and further, that it did provide him guidance for thinking about a problem met subsequently. He appreciated his Logo shape families experience as embodying a powerful idea.

Did he apply the heuristic because it might generate new results or because it might help him understand a range of puzzling phenomena? We can not profitably make such a distinction if his purposes were mixed. The two aspects of purpose *we* might choose to distinguish appear to have been inextricable *for him.* He used this powerful idea as a heuristic for orderly exploration to generate interesting results in a comprehensible way.

We have come to the end of this story, but the question remains 'where does this powerful idea of variable stepping go from here?' We find a hint in the discussion of Robby's new theory. His conjecture, that the odd-numbered chains make two separate figures, was based on the regularity he observed in five cases (with two through six rings).

When I asked him to prove this new theory, his method of choice was empirical — he constructed a seven-ring puzzle and cut each of the rings. (I did not suggest his doing so. I wanted him to reflect more.) He clearly expected the seven rings to separate into two figures and took their doing so as proof of his theory. *What is significant in this observation is the way hypothesis testing emerged as a minor variation from a preceding activity which was a theory-free but orderly exploration of an interesting domain.* Variable steping had become for Robby a way of approaching the world, of seeing 'what's what'. *The power of the idea* — as witnessed by Robby's quick invention of his new theory — *is that from 'what's what' 'what follows' is 'intuitively obvious'.*

ACKNOWLEDGEMENT

A discussion of this material with Gay Drescher helped to clarify some issues touched upon in this chapter.

REFERENCES FOR PART I

Barker, R. and Wright, H. (1971) *Midwest and its Children.* Hamden, Ct.: Archon Books (original edition 1955).

Fang, J. (1970) *Towards a Philosophy of Modern Mathematics.* Hauppauge, New York: Paideia series in modern mathematics, Vol. 1.

Feynman, Richard. (1965) 'New textbooks for the "new" mathematics'. In *Engineering and Science,* March. Pasadena: California Institute of Technology. Also see his further comments in the autobiography, *Surely You're Joking, Mr. Feynman.* New York: W. W. Norton, 1985.

Goodall, J. (1971) *In the Shadow of Man.* Boston: Houghton Mifflin.

Inhelder, B. and Piaget, J. (1958) Flexibility and the operations mediating the separation of variables', a chapter in *The Growth of Logical Thinking from Childhood to Adolescence,* translated by A. Parsons and S. Milgram. New York: Basic Books.

Langer, S. (1948), *Philosophy in a New Key.* New York: Mentor Books, The New American Library.

Langer, S. (1967) *Mind: An Eassy on Feeling,* Vol. 1. Baltimore, Md.: The Johns Hopkins Press. See especially Chapter 5, 'The idols of the laboratory'.

Lawler, R. (1979) *One Child's Learning.* Unpublished MIT doctoral dissertation.

Lawler, R. (1981–82) 'Logo ideas', a column appearing in *Creative Computing.*

Lawler, R. (1983) 'Designing computer based microworlds'. In M. Yazdani (ed.) *New Horizons in Educational Computing,* Chichester, England: Ellis Horwood.

Lawler, R. (1985) *Computer Experience and Cognitive Development.* Chichester, England: Ellis Horwood.

Lévi-Strauss, C. (1966) *The Savage Mind.* Chicago: University Press.

Minsky, Marvin (1969) 'Form and content in computer science'. ACM Turing Lecture, August 1969; published in the *Journal of the Association of Computing Machinery,* April. Also available as Memo No. 187 from the MIT Artificial Intelligence Laboratory.

Newell, A. and Simon, H. (1972) *Human Problem Solving*. Englewood Cliffs, N.J.: Prentice-Hall.

Papert, Seymour (1971A) 'Teaching children to be mathematicians versus teaching about mathematics'. In *International Journal of Mathematical Education in Science and Technology*. New York: John Wiley, 1972. Papert's papers are available from the MIT Arts and Media Technology Center.

Papert, Seymour (1971b) 'Teaching children thinking'. In *Mathematics Teaching*. Leicester, UK: The Association of Teachers of Mathematics, 1972.

Papert, Seymour (1973) 'Uses of technology to enhance education'.

Papert, S. (1980) *Mindsorms: Children, Computers, and Powerful Ideas*. New York: Basic Books.

Piaget, Jean (1953) 'How children form mathematical concepts'. *Scientific American,* November, reprint 420. San Franciso: W. H. Freeman.

Piaget, J. (1971) *Biology and Knowledge*. Chicago: University of Chicago Press.

Piaget, J and Inhelder, B. (1975) 'The development of operations of combinations,' a chapter in *The Origin of the Idea of Chance in Children*. New York: W. W. Norton.

Simon, H. (1969) *The Sciences of the Artificial*. Cambridge, Mass: MIT Press.

Part II

Logo confessions

Benedict du Boulay

The chapters in this part of the book describe the use of Logo with student teachers in an experiment going as far back as the mid-1970s. This started as work towards a Ph.D. and was later expanded into a research project funded by the then British Social Science Research Council. The aims were to explore the teaching of Logo in a mathematical context to student teachers in mathematical difficulty and to try to distinguish the reality from the rhetoric surrounding the language. The research was part of a more general effort by a team of people in the Department of Artificial Intelligence, University of Edinburgh, who were working mostly with children and exploring various aspects of Logo in education. My objective in this Part is to give the flavour of the work with student teachers through case-studies of three individual students. These provide a 'warts and all' account of what teaching and learning Logo was like. I hope that the other contributions in this book and the more widespread knowledge of Logo will provide enough of a context and so I have omitted both a description of the Logo language itself and a survey of other Logo-related research.

When the study was started in 1976 there were no microcomputers in schools or Colleges of Education. Our Logo system was implemented on an ICL mainframe and needed a separate minicomputer with ferrite-cored memory and racks of electronics to drive the drawing devices which consisted of graph-plotters, a storage-tube display and a mechanical (floor) Turtle. This latter was improvised out of Meccano, there being none to be bought at the time. The students had to type on clackety model 33 Teletypes and the Logo laboratory was a noisy room festooned with cables. Logo was still a rather special language and the only institutions in the UK which had access to it were those taking part in research programmes

I have called my section of the book 'Logo confessions' for several reasons. First it is a reaction to publishing data from my thesis some years after the event, a sort of academic coming out. More importantly, the students that I worked with sometimes seemed to regard the Logo sessions as periods where they could safely reveal their mathematical sins and inadequacies for discussion (and in some cases absolution). The religious sentiment amused me and fitted in rather well with the apostolic fervour that seems to grip purveyors of new programming languages as well as with the departmental culture influenced by lapsed-Catholic, Guinness-drinking readers of Myles na Gopaleen who were around in Edinburgh at that time. One of the pieces by Fr. Hacker, an occasional contributor to the more esoteric UK Artificial Intelligence and Expert systems newsletters, captured the flavour exactly. He taught us our catechism . . . What is Logo? Logo is the one true programming language. Who made Logo? Papert made Logo . . . and so on. Much to my surprise I recently received a little pamphlet which adopted the Hacker approach as a serious marketing strategy.

ACKNOWLEDGEMENTS

I would like to thank the very many people who helped me complete this research. Jim Howe, my supervisor, gave me the opportunity to start it and guided me through it. Many of the ideas came from discussions with Tim O'Shea, to whom I am much indebted. Members of the Department of Artificial Intelligence provided a stimulating environment in which to learn how to undertake research. Colin McArthur wrote and maintained Logo, and John Kemplay helped build and maintain equipment. Ailsa Anderson assisted at some Logo sessions.

Frazer Paxton, Robert Finlayson, Ian Lusk and other staff members of Moray House College of Education allowed me access to their students, lent me equipment, advised me and gave me introductions to head-teachers. The head-teachers of many primary schools in the Lothian Region and in Fife gave me permission to observe student teachers while they were on teaching practice. Without the willing cooperation of the anonymous student teachers there would have been no research. The Social Science Research Council supported me financially.

My greatest debt is to my wife and children who put up for so long with someone who was 'doing his thesis'.

5

Setting The Scene

WHY LOGO?

At the time of this work Papert's *Mindstorms* had still to be written, though many of the MIT Research Papers and Technical Memoranda on which the ideas are based were available. In some ways the starting point for the work was the following quotation from Papert (1973):

> Most people emerge from high school without ever having had a joyful or personally meaningful mathematical experience. No wonder they hate it and refuse to learn it! We think it is important and *easy* to remedy this for college students in academic trouble, for future teachers (especially) . . . . (p. 36, his emphasis)

I wanted to see whether student teachers, destined for primary schools and suffering from various difficulties in mathematics, could be helped by a course in Logo individually aimed at their particular problems. There was some evidence at the time that a proportion of the students graduating from Colleges of Education were not adequately equipped to teach mathematics effectively. Given the failure of traditional methods with this kind of student, it seemed sensible that an alternative approach based on Logo should be tested.

There were four kinds of claim about the relation of learning Logo to learning mathematics. These were:

(1) Programming provides some justification for, and illustration of, formal mathematical rigour.
(2) Programming enables mathematics to be studied through exploratory activity.
(3) Programming gives insight into key mathematical concepts.

(4) Programming provides a context for problem solving and a language with which the students may describe their own problem solving.

My objective was to examine each of these claims.

THE EXPERIMENT

A number of students were recruited as volunteers from Edinburgh College of Education where they were taking a three year Diploma of Education. As part of the experiment they were taught Logo, given mathematics and attitude tests, observed on teaching practice, involved in individual discussions about mathematics and given Logo-based tasks related to their mathematical needs. The aims of the testing, observation and discussion were partly to determine what the students' mathematical difficulties were and partly to try to assess the effects of the Logo-based intervention in their training.

The study observed fifteen students, of whom three, Jane, Irene and Mary, were dealt with in greater detail. Various issues were examined including the students' mathematical difficulties, their progress (or lack of it) in learning to program and the interaction between their Logo work and their mathematical understanding. It is the data from the last of these that is described here, namely how their work in Logo interacted with their mathematics. Three mathematical case studies are given, derived from notes, tape-recording and records of their interactions with the computer. No attempt is made here to describe the Logo implementations that were used, nor to indicate what parts of Logo turned out to be either easy or hard for these students to understand. Further details of the work can be found in du Boulay (1978) and du Boulay (1980).

As volunteers, all three students believed (rightly) that they had difficulties with mathematics and they hoped that the experiment would help them in their future day-to-day work as primary school teachers. In the event their reactions to the experiment were very different. Each, I believe, benefited from having someone, outside the College of Education, who was (usually) sympathetic and willing to listen and to try to help. Students had surprisingly few opportunities and little inclination to be honest with their College lecturers about their misconceptions and fears. For example, observation by the lecturers during teaching practice seemed, to the students, to be as much about assessement as about help and advice. My own intrusions into these sometimes tense and difficult lessons were, I like to think, viewed differently, even if they were just as nerve-racking for the students, as an extension of the Logo work: a chance to make ideas explicit, especially if it meant isolating and exploring some area of difficulty.

The students attended the Logo laboratory at the University on a semi-regular basis after lectures at the College of Education. All the students undertook a general-purpose introductory course in Logo programming for a term before their first teaching practice. Each year students spent two terms in College and one on teaching practice in local primary schools.

During the first teaching practice there were no Logo sessions. Rather I visited the schools where they were working to observe and record selected mathematics lessons and to discuss with them how these lessons had progressed. Difficulties thus identified were later used as the basis of individual Logo-based work. Jane, Irene and Mary continued to attend the laboratory for four more terms including another term of teaching practice.

The general pattern was of small amounts of Logo work spread out over a long period. This was not ideal but the only method that was possible in the circumstances. In retrospect, a more concentrated and intensive course might have been more effective if it could have been organized.

Questionnaires and tests were administered at various points to help assess changes in attitude and skill. These are referred to at various points in the case studies. Reference is also made to a change in the implementation of Logo part-way through the experiment. This gave us a much nicer Logo system but the small changes in behaviour of the system did upset some of the students for a time.

One other piece of equipment, in addition to that mentioned in the introduction, was also available in the latter stages of the work. This was a Button Box used to drive the turtle. The box had sixteen buttons and could be held in the hand. Some buttons would drive the turtle directly, others could be used to define procedures.

A rather mixed picture emerges from the case studies. There were clearly some gains for the students, some ideas now understood, some sense of personal discovery—but there were costs. For example, Irene found the experience just as unpleasant as traditional mathematics had been and some issues turned out to be very hard to explore given the available computational tools.

ADVANTAGES AND DISADVANTAGES OF LOGO

Learning mathematics through programming has both advantages and disadvantages, which are summarized in Table 5.1. They are now described against the background of the claims commonly made for this approach to learning mathematics (see, for example, Feurzeig *et al.* (1969)).

Rigour and exploration

The work given to the students did not attempt to illustrate the idea of mathematical rigour (e.g. as applied in proofs) but succeeded in demonstrating the value of the weaker idea of 'explicitness'. Students realized that the lack of ambiguity in the programming language enabled them to communicate their intentions to the computer exactly. But this was often a frustrating experience. Some students did not find the required precision of expression congenial and compared the effort involved unfavourably with the result (e.g. an hour or so spent drawing a simple house outline).

Programming provided extensive opportunities for the students to engage in active mathematical exploration, particularly of Turtle Geometry. Mary made several personal mathematical discoveries by investigating the

Table 5.1.
Advantages and disadvantages of learning mathematics through programming.

Advantages	Disadvantages
Emphasizes importance of explicit language.	But causes frustration because making intentions clear can take a long time.
Provides opportunities for active mathematical exploration, e.g. Turtle Geometry.	But primitives for other domains are too complex for students to write for themselves. Students may work at the wrong level of representation.
Certain key concepts are well illustrated, e.g. function algorithm, angle, state and transformation.	But availability of the computer distorts syllabus towards what is easy to program rather than what the student needs. It is not easy to design programming projects which address the mathematical difficulties of students.
Splendid opportunities for problem solving are available.	But ease of access to the machine encourages trial and error rather than analysis.
Learning programming through Turtle Geometry is fairly effective and fun.	But certain concepts still difficult, e.g. variables. Strategies evolved in solving drawing problems inappropriate for other classes of problem. Some students find programming just as frightening as mathematics.
Successful completion of projects improves attitude to the topic studied.	But overall attitudes to mathematics hard to change. Student teachers often more concerned with recipes for successful teaching than with understanding the topics they are to teach.

mathematics underlying her programming. Different primitives were needed for other domains. These turned out to be too complex for the students to construct for themselves out of Logo and were provided for them. This happened to an increasing extent towards the end of the study. It freed the students from the need to worry about tedious and often irrelevant programming detail and enabled them to concentrate on the mathematical properties of the given new primitives. For example, primitives were provided to illustrate the symmetry transformations of the rectangle and to give a visual interpretation of fraction operations based on ratios. In both cases the main programming effort concerned the method of mapping from the internal representation to the drawings on the display screen. This had scant mathematical value and would have been an inappropriate task for these students, even if they had the programming skill to undertake it. The students concentrated on the way the primitives behaved rather than on the way they were constructed. This did not mean that they did not think carefully about the behaviour. For instance, Jane found a mathematical bug in one of the primitives.

Key concepts

Programming illustrated a number of key concepts including function, algorithm, the state/transformation distinction and angle as rotation. These concepts were embodied in the structure of Logo. Sometimes students explored a concept by observing how the primitive behaved (e.g. what it drew). By contrast, Mary was able to use the code of hypothetical procedures to learn about functions. But the availability of the computer tended to distort the students' work towards what was programmable rather than what was mathematically most beneficial. For example, it was found in many cases that students knew certain algorithms (which were easy to program) but they did not understand the meaning of these algorithms or how they might justify and explain them to their pupils. This made the reproduction of the algorithms in a program a pointless activity.

Problem solving

The students had splendid opportunities for problem solving. Mary and Jane used this to learn the value of the strategy of problem decomposition. But the ease of access to the machine encouraged trial and error or bottom-up methods rather than planning and analysis. For instance, certain figures were drawn without reference to their overall geometric properties. Instead the student used bottom-up 'incremental' strategies which depended only on detailed low-level analysis. Most students thought about their programming activity and the way they had learned to program. Mary and Jane were able to make some analogies between their experience of learning programming and children's experience of learning mathematics.

Attitudes

Most students enjoyed learning to program using Turtle Geometry but found certain concepts hard to grasp, e.g. variables. The bottom-up problem-solving strategies which the students evolved were not appropriate for other classes of problem, e.g. symbol-manipulation problems and decomposition into functions. Some students, notably Irene, found programming difficult and just as frightening and unpleasant as mathematics had been.

Some small improvements in attitude to mathematics were observed. These were linked to the particular mathematical topics which had been explored. The students' overall self-confidence and attitude to mathematics was not changed (and seemed very hard to change). Many of the students held the pragmatic view that their main task was to pass their Diploma and learn a number of 'recipes' for teaching mathematical topics. For them, learning programming in order to understand mathematics better seemed a circuitous route to more successful teaching. Differences of initial attitude were important. Mary seemed to believe that her mathematical difficulty consisted of ignorance or lack of understanding of specific topics. These she regarded as separate, solvable sub-problems which could be tackled individually. By contrast, Jane imbued her ignorance and lack of understanding with a general belief in her own mathematical incompetence. Although she became more confident about teaching specific topics, her overall attitude remained unchanged.

It was found that students needed explicit help in linking the mathematical content of their programming projects to their existing mathematical knowledge. For example, Jane still had difficulty understanding clockwise and anti-clockwise rotations on a protractor after extensive work in Turtle Geometry. Only after I linked the turtle rotations specifically to the protractor did she see the connection.

The two major disadvantages of learning mathematics through programming were that programming itself was a complex skill for these students to learn and that it was all too easy to give the students problems to solve at the wrong level of representation. For example, some students were asked to study fractions by writing procedures to draw fraction pie-charts. This was a poorly chosen project because most of the students' efforts were directed at getting the drawing correct. This could be achieved (or could fail) with little reference to fractions and did not help their understanding of fractions. In this project, and in another on vectors, the students were asked to give commands which drew representations of the given mathematical structures. This was not the same task as giving commands to manipulate an internal representation of those structures, where the manipulation had the side effect of producing a drawing. The latter task would have been more valuable because it involved understanding the structures themselves rather than the *pictorial* qualities of their representation. In general, it was not easy to design programming projects which helped the students deal effectively with their classroom difficulties. This was because the programming was too complex, or because the behaviour of the program or its code did not demonstrate the idea in question clearly.

The main advantage of programming was that it presented mathematics as an exploratory activity involving problem-solving rather than rote learning. It provided concrete illustrations of a number of abstract ideas. Finally, it gave the students the opportunity to be honest about their mathematical difficulties and to discover that they had the ability to overcome at least some of them.

6

Jane

Jane spend about 84 hours in the programming laboratory writing and debugging programs as well as discussing mathematics. This does not include the time spent discussing recordings of her lessons or the time spent in meetings at the College of Education. The number of hours Jane spent programming, month by month, is shown in Fig. 6.1. In all, Jane took part in

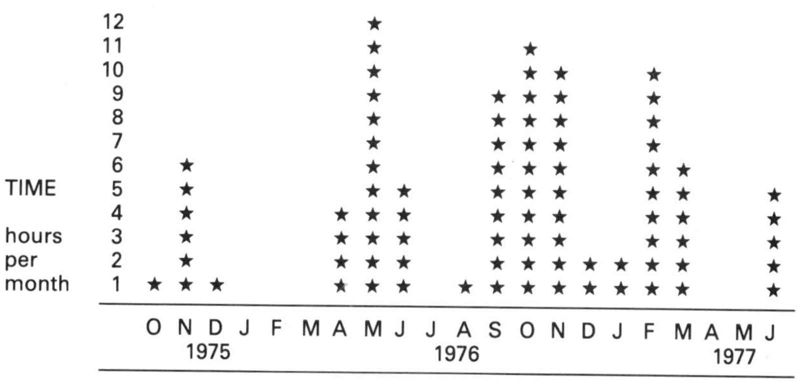

Fig. 6.1 — Jane's programming sessions.

the project over a period of 22 months. The three monthly vertical divisions in Fig. 6.1 correspond roughly to College of Education terms.

Jane's work will be presented by concentrating on the evolution of her understanding of particular topics, such as angles. In this way it is intended to trace the course of the interaction between her programming and her mathematics.

The next three sections analyse her work in the areas of geometry, algebra, and number. A final section summarizes her work in relation to the framework set out earlier.

GEOMETRY

Jane had a number of misunderstandings about geometry both at the start of this study and during the course of it, e.g. use of a protractor. This section describes the programming work she undertook and the effects it had on her understanding of such concepts as angle. Angle as rotation is a central concept in Turtle Geometry. It will be shown how Jane's attitude to, and skill in dealing with, angles improved as a result of her programming work. As her work progressed it will be shown how her attention shifted from the properties of particular angles, with which she had become familiar, to the more general properties of shapes, for example their symmetry and their total angle properties.

The importance of good teaching will also be seen, especially where opportunities were lost for exploiting Jane's programming work for mathematical purposes. Evidence will also be presented which suggests that despite her extensive angle work in the programming classroom, Jane failed in some respects to link it with her existing angle knowledge. Instances will also be given of how concentration on the programming aspects of a problem sometimes masked what was important mathematically.

Learning to visualize angles

The elements of programming were learnt, in the first term, by writing procedures to control the floor turtle or the simulated turtles in the display and graph-plotters. Jane succeeded in drawing a number of designs including regular polygons. This necessitated that she distinguish translation from rotations, and clockwise rotations from anti-clockwise rotations. In order to draw a given polygon it was necessary for her to know the values of its angles in degrees and to distinguish interior from exterior angles of the polygon. Fig. 6.2(a) illustrates our use of the term 'exterior' angle of a polygon.

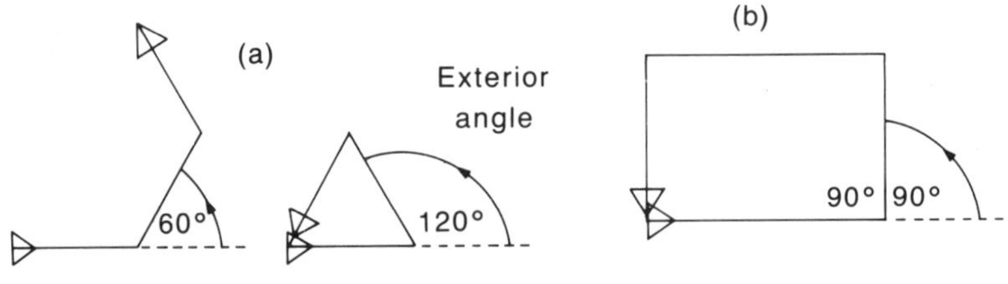

Fig. 6.2 — Interior and exterior angles.

Occasionally she failed to make this distinction and produced, for example, part of a regular hexagon instead of an equilateral triangle, see Fig. 6.2(a). This was a common mistake among the students, probably accounted for by the misapplication of the rule that 'the three angles of a triangle add up to 180 degrees'. It may also have been affected by drawing squares and rectangles whose interior and exterior angles are both 90 degrees (see Fig.

6.2(b)) and whose successful completion did not confront the student with this interior/exterior angle distinction.

When Jane returned to programming, after teaching practice, she made just the same mistake when drawing an equilateral triangle, as part of the problem of drawing a 'house'. Her initial diagnosis was that she had got the direction of the rotation wrong rather than its magnitude:

> 'I've done it the wrong way. I said right instead of left.'
> '...instead of saying left sixty, I said right sixty.'

I asked whether a triangle could be drawn by turning right. Jane agreed that this was possible and then sorted out the difficulty on her own, as follows. She next tried a rotation of RIGHT 150, but changed this to LEFT 120 when she saw its effect. She was asked what the trouble with her procedure had been and replied:

> 'Well, em, I should have done a hundred and twenty plus, a hundred and twenty round... you see six... I always get this wrong. I keep forgetting that the point's [the light-spot representing the turtle on the display] going that way and therefore you have got to get that far round. I think it... I always think it's like a protractor... that... I was measuring sixty but we should be have been six... *ninety plus thirty*, that's one hundred and twenty.'

The value of programming in this context is clear. Jane was able to solve a problem herself because she was able to exploit the link between her explicit commands and the machine's visible reactions to those commands. The separation, in Logo, between procedures for producing rotations and those for producing translations is valuable in this respect. However, her explanation of her difficulty and its solution, particularly the phrase 'ninety plus thirty', suggest that she was not using the idea that the interior and exterior angles are supplementary (add up to 180 degrees). She seems to have solved the problem by a process of successive refinement, after her initial attempt. This problem was to recur with her and other students. The ease of access of the machine on which experiments could be conducted tended to prevent the students conducting a rigorous analysis. Very often local problems could be solved by trying solutions which were gradually refined. However, once the solution had been found, there was little incentive to examine it, or the process leading to it, to find other solutions.

Jane then spent some time fitting the triangle correctly on top of the square to finally complete her 'house', see Fig. 6.3. It took quite a lot of time and effort, partly in dealing with procedure management (e.g. defining, editing, listing and saving procedures) and the syntax of the language, and partly in working out how much to rotate the turtle between drawing the triangle and the square. Difficulties with the language were aggravated by a change of Logo implementation. When it was done she explained:

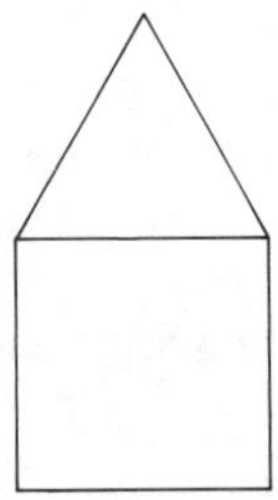

Fig. 6.3 — House.

> 'Oh, huh, its an *awful lot of paper just to do that*. I suppose you learn by... *by making mistakes and having to put it all together again... I think you do*. It really makes you think about it. I think what really stumps me is the fact that I can never remember which way it's [the simulated turtle] facing. *Like to make the angles, I completely forget which way, like it is going that way and it isn't like a protractor and facing up the way*.'

Part of her difficulty was caused by the fact that the current heading of the simulated turtle on the storage-tube display was not shown continuously. It was only possible to request that it be shown briefly. I suggested that she did more work with the floor turtle whose heading was easier to see.

Her phrase 'just to do that', that is draw a simple house, indicated that in terms of a *product* the session had been hard work for a small return. But she qualified this with her remarks about learning from mistakes which suggested that she saw value in the *process*. That is to say, the rigour demanded by the computer was seen as valuable because it forced her to formulate a completely explicit description of the 'house'. It also suggested that she was looking on her mistakes constructively and was using the activity to 'think about thinking', though without using computational terms to describe her thinking. By virtue of this explicitness, she could debug her procedures. But as we have noted, there was little overall analysis of the problem. This disadvantage of programming might be diminished if students were directed through a more formal procedure planning stage than was the case in this study.

I lost the opportunity for making Jane be more verbally explicit about the interior/exterior angle distinction. I just accepted her phrase 'it isn't like a protractor' which I understood to refer to the interior angle. It seems clear that, whatever potential the programming activity has as a method of studying mathematics, skilful teaching still has a vital part to play. Here the explicit computational description of the square and triangle could well have been complemented by an overall analysis of the solution found and by an explicit English description of their properties and interactions. Just because Jane had succeeded in solving the current problem did not necessarily mean that she had understood her solution.

At the next session, she used the house procedure to draw a whole 'street' of 'houses'. When this was drawn on the display screen it came out upside-down, because of the initial heading of the turtle at the start of the drawing.This could have been cured by initially rotating the turtle 180 degrees. Instead Jane wondered:

> 'Would that be three... that would be right.. that would be right round, er, wouldn't it.'

I said that turning the turtle right round would make no difference, but the problem was left hanging to be solved later. Jane seems to have muddled turning something upside-down, rotating it through 180 degrees, and turning something right round, rotating it through 360 degrees. Again a mathematically interesting point was passed over because, in some sense, the programming problem had already been solved. A 'house' had been drawn successfully. It was easier to turn the paper upside-down with the house drawn on it than to worry about why the house came out upside-down.

Some of the questions in the second questionnaire, administered that term, concerned the planning of drawing procedures. The students were asked how they would set about producing the pattern given in Fig. 6.4. Jane

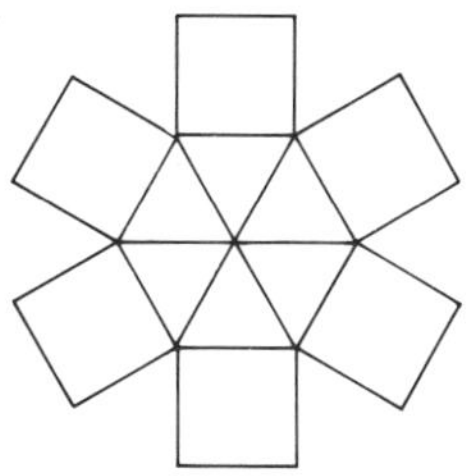

Fig. 6.4 — Rotated house.

gave a clear English description, breaking down the problem into appropriate sub-problems, of which drawing a house shape was one. She correctly explained that each house would have to be rotated through 60 degrees in relation to its neighbour.

Another question asked if she could see any use for her programming work in her future teaching. She replied:

> 'I've a much better idea of angles degrees etc. than I had when I started — *before I couldn't have attempted it without a lot of help*.'

An important phrase here is 'before I couldn't have attempted it'. It suggests a positive change of attitude about her ability to tackle this topic in the classroom which may be contrasted with her normal lack of mathematical self-confidence.

At a later session that term Jane wanted to draw a rhombus which she called a diamond. At first she wrote a procedure in which all the angles were 45 degrees that produced a drawing as in Fig. 6.5(a). She then changed all

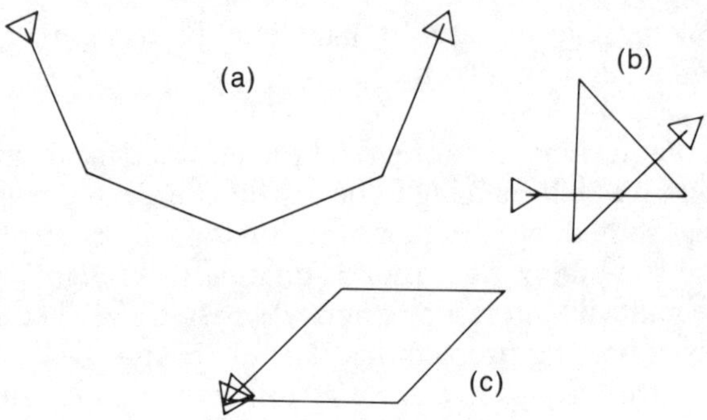

Fig. 6.5 — Drawing a rhombus.

the angles to 135 degrees which produced a drawing as in Fig. 6.5(b). This suggests that she was still confused about the interior/exterior angle distinction but knew they were supplementary. She sorted out the problem on her own to produce a non-state-transparent rhombus with angles 135, 45 and 135 degrees, as Fig. 6.5(c). A state-transparent figure leaves the turtle with the same heading and position as at the start of the figure. She tried to use this rhombus to draw a 'flower figure' as in Fig. 6.6, but found difficulty in

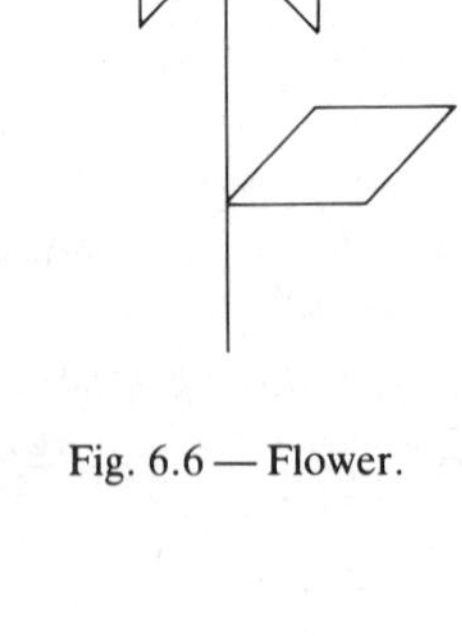

Fig. 6.6 — Flower.

aligning the turtle between each rhombus, no doubt because of the lack of state-transparency. She did not seem to have made an overall analysis of the problem, as she had done in her answer to the questionnaire question about rotated houses, and was proceeding bit by bit to build the flower.

Drawing polygons

Later in the term Jane gained further experience of the angle properties of polygons. The sessions was designed to enable Jane to explore angle properties of regular polygons and to consolidate previous work with user-defined procedures which took arguments. She was shown how to define a procedure of two arguments which could draw any regular polygon depending on the values of the argument (one for the exterior angle, the other for the number of sides). She was introduced to the term 'regular' and after some initial work with convex polygons, it was suggested she try drawing 'stars'.

She tried to draw a star by choosing various argument values for the procedure but could not, at first, make the ends join up, see Fig. 6.7. She had

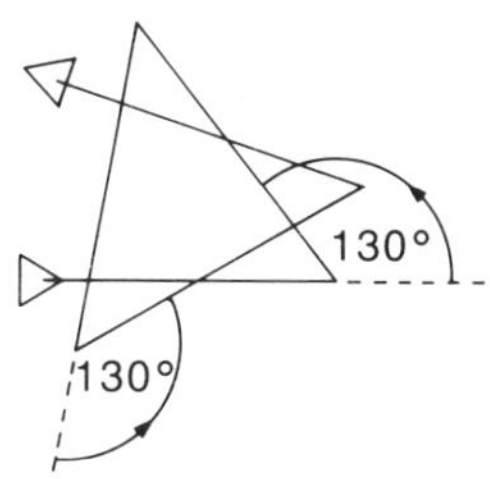

Fig. 6.7 — Failed star.

not understood that the condition for this to happen was that the turtle rotate a whole number of complete revolutions in tracing the star. This is known as 'the total turtle trip thereom'. Discussion revealed that she did understand the relation between the sharpness of the star's vertices and the angle argument of the procedure. That is that 150 degree would produce a more sharply pointed star than 130 degrees. Eventually she was able to draw a five-pointed star by trial and error. Her last two attempts used angles of 145 and 143 degrees. Within the limits of accuracy of the display these could not be distinguished from the star with the correct angle of 144 degrees. She was asked if she knew any rule which could help her to draw stars but she said not. She then spent some time modifying her polygon procedure, by adding an extra variable, so that polygons of any size could be drawn. She drew a number of decagons of different sizes and understood that these closed because she had specified a complete revolution in ten individual turns of 36 degrees each, making a total of 360 degrees in all.

She then returned to the problem of the stars, but could not see the relevance of the above condition of closure of the decagon, especially as the total angle turned in drawing her stars was much greater than 360 degrees. This difficulty was aggravated because she was certain that her five-pointed star, using 143 degree angles, had in fact closed. Thus when she was asked to calculate its total angle turned, she correctly calculated 5*143=715 degrees but saw no significance in this result.

Jane was asked to imagine herself walking around the star, but in the end I specifically pointed out that exactly two complete revolutions closed the figure. The dialogue became very one-sided, as I explained to Jane. At first Jane thought that failure to close was because she had failed to arrange for her procedure to turn the final corner of the star, but this was not the case. She was near the truth because she was using the idea of state-transparency in her analysis of the problem, but was not using it correctly in this case. Eventually she understood, after further intervention by me, that a five-pointed star needed a total turn of 720 degrees. That is two complete revolutions made up of five rotations of 144 degrees. Jane explained her difficulty:

> 'I didn't realize that with a star. You know, I never thought in terms of going round several times. I was always thinking in terms of one revolution. Always it was three sixty.'

Unfortunately the opportunity was not grasped to suggest that she draw other stars and the session moved on to a new topic. This incident shows up one practical difficulty associated with the particular drawing devices in use. It also shows how both of us were over-concerned with the product, that is in this case a star, so that insufficient time was spent on studying the general properties of stars. Perhaps if I had not pushed Jane for an explanation of her first star, the total turtle trip theorem would have emerged more naturally from Jane's own exploration of the problem of drawing stars with different numbers of vertices. As it was, I gave her 'the answer'. A further difficulty was caused by the vagueness of the instructions, 'try to draw a star'. It would have been better to suggest she drew a number of stars possibly with specified numbers of vertices. At the end of the session, which had also included some work on functions, Jane was asked if she felt her programming work was helpful. She replied:

> 'Well, yes, I think so because it really makes you think about, you know, how you do these procedures... Yes, well, I think with the angles... I think it's, I think it's probably getting a bit easier now to think about angles. Whereas before it was just a, a haze. I didn't have a clue... I... usually I am absolutely terrified in case any of the children in the school had asked me about angles because I was just... you know, I would have to go away and really think about it and then come back. But if they want an answer right there and then, I just used to have to fob them off with something else.'

Note her fear of beng found out not knowing the answer to a child's question. She is still rather guarded in her evaluation, 'a *bit* easier now to think about angle'. Her statement suggests a change of attitude because she believed herself to be now more capable of tackling this topic in the classroom. This belief in her ability to cope mathematically is important for her success and happiness as a mathematics teacher. Of course, mere belief is not enough, it must be underpinned by understanding; but it remains an important factor.

Looking for patterns

Jane continued to program during the summer vacation. This was partly because she enjoyed it and also because she did not want too long a gap in her work. The last time this had occurred, during the previous teaching practice, she felt that she had had trouble revising programming.

One of the problems she worked on was to draw a 'tree' using an 'arrow' sub-procedure, see Fig. 6.8. During the course of this problem, Jane

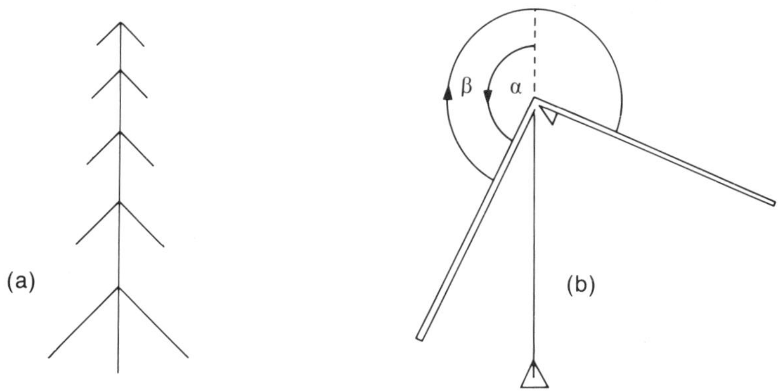

Fig. 6.8 — Tree and arrow.

explored how considerations of symmetry enabled her to deduce the angle properties of a figure. Her analysis was initiated as a way of reducing inefficient computation. She was trying to construct a mathematical rule for herself. She defined an arrow procedure with arguments which allowed her to vary the size of the arrow's arms and spine, as well as the inclination of the arms to the spine, see Fig. 6.8(b). In the figure, the angles which affected the inclination are marked alpha and beta. She wished to draw an arrow whose arms were at equal inclinations. After a calculation, described presently, she tried angles of 135 and 215 degrees which gave her an arrow similar to that in Fig. 6.8(b). She then recalculated and tried 135 and 270 degrees which gave the required effect. She realized that the symmetry of the arrow would force a numerical relation on these two angle values but did not know what that relation was.

She asked for help, explaining that she did not want to redo her (laborious) calculations each time she wanted to draw an arrow with arms at equal but new inclinations to the spine. She knew that having selected the value of one angle, then the value of the other angle was determined, but did not know how to find it. This provides an example of the generation of a search for a mathematical abstraction by the need to solve a concrete problem more efficiently. Strangely, Jane was not alerted to the relation (that one angle should be twice the other (see Fig. 6.8(b)) by the values 135 and 270 degrees. This time I suggested that she solve this problem for herself.

By the next session, she had solved the problem and understood that one angle had to be twice the other. She explained that previously she had calculated the values by breaking up the rotation into parts (see Fig. 6.9) and

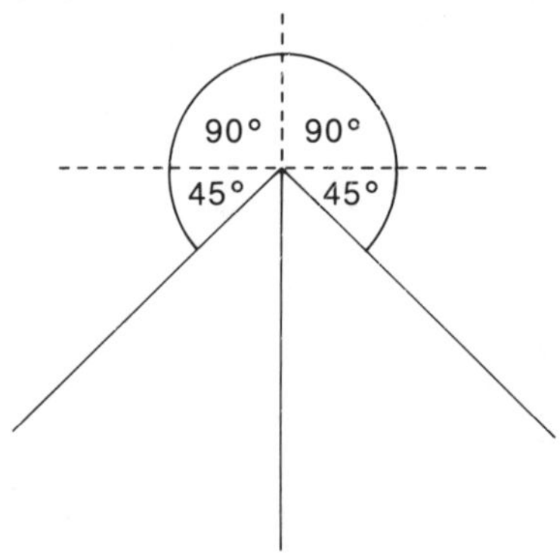

Fig. 6.9 — Arms of arrow.

then adding those parts. It had been a mistake in this addition which had made her use 135 and 215 degrees instead of 135 and 270 degrees. She also explained that she had:

> 'looked and looked [at 135 and 270] and thought there was no relationship at all... [because]... I said twice one-thirty-five is *three* seventy.'

This suggests that she might well have suspected what the relationship was but was misled by an arithmetic error. Similarly in the incident of drawing the star, earlier, she did not notice that 715 was approximately equal to 720 (2∗360). Having successfully mastered the arrow problem she went on to draw a 'wood' full of 'trees' made up of 'arrows'. This incident supports the claim that programming provides an activity in which mathematical questions can be tackled. Here the mathematics was employed by Jane to make a problem solution more elegant. But it also indicates that the activity should be assisted by sympathetic teaching.

In a later session Jane explored a number of different recursive procedures including ones which printed out a Fibonacci series (see section 6.3) and another which generated a number of spirals. The spirals procedure was a generalization of the polygon procedure she had already worked with. The spiral was produced by drawing line segments of increasing length whose angle of inclination to each other was fixed, see Fig. 6.10. This gave her the

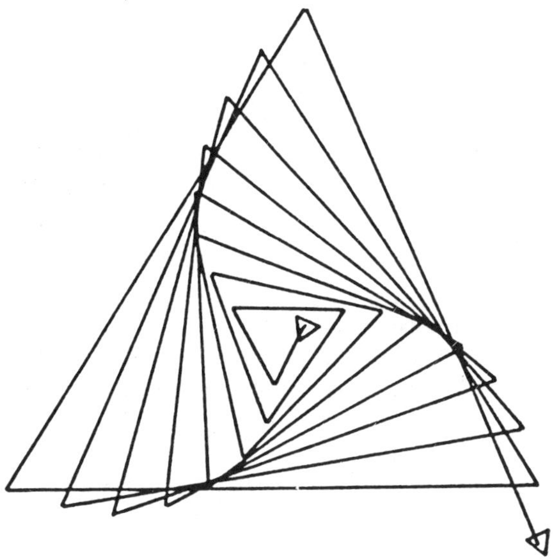

Fig. 6.10 — Spiral.

opportunity to explore the properties of obtuse and reflex angles. She worked on the problem of reproducing the sharply pointed spiral of Fig. 6.11 which was also on the wall of the programming classroom. This spiral had an exterior angle of 170 degrees. The task here was not to write the procedure, which was given in the worksheet, but to find appropriate arguments for it. She tried angles of 135 degrees, then 200 degrees and then 250 degrees. In the process she did not notice that, by passing through 180 degrees, she had inadvertently changed the sense of the sprial from clockwise to anti-clockwise and was not making the vertices blunter by increasing the angle. I suggested that she try drawing spirals of 90, 180, 270 and 360 degrees which she did. Angles of 90 and 270 degrees produced 'square' spirals winding in opposite senses. While angles of 180 and 360 degrees were degenerate cases, producing an oscillating, expanding line and a lengthening line respectively.

When she returned to the problem of the sharply pointed spiral, she was able to suggest 160 degrees as a good value to try because it nearly sent the

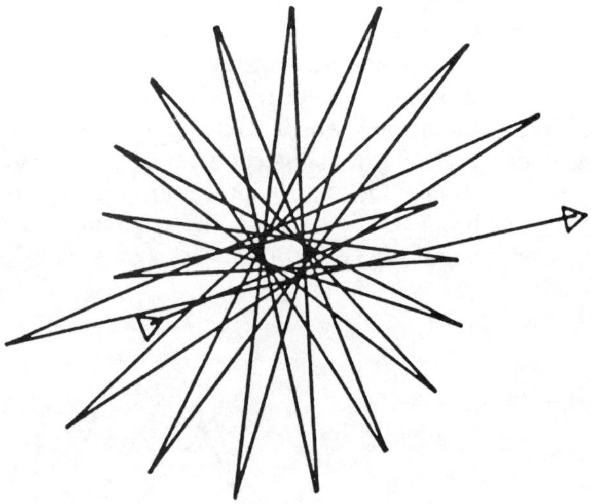

Fig. 6.11 — Sharply pointed spiral.

turtle back on itself. By small increments on the initial value of 160 degrees she ended up with a spiral, which she liked, with an angle of 170 degrees. This provides an instance where the reactive nature of the computer allowed her to explore angle properties and at the same time produce pleasant designs (in contrast to the rather unexciting 'house', described earlier). In this case she had spent a lot of time previously establishing how the recursive spiral procedure produced these effects, and in this session she was free to just run her spiral procedure with different angle argument values to observe the effects.

Both her exploration of the symmetry of the arrow and the sharpness of the spiral vertices were the result of her attempting the solution of some other larger problem, though it must be mentioned that both the work with 'trees' and with 'spirals' was suggested in the worksheets.

Linking turtle and school geometry

The following term, Jane was on teaching practice again. She taught two lessons in which she was observed giving the children practice in the use of ruler, compass and protractor to construct and measure triangles. In a programming session between these two lessons, her difficulties in the use of a protractor were observed.

These lessons and the intervening programming session will now be described because they show how Jane had failed to link her knowledge of angles as rotations (turtle geometry) to the use of the protractor (school geometry). Only when I had pointed out the link to her, was she able to make use of it in the way she taught the children.

The first of the observed lessons concerned the naming and construction of different classes of triangle, e.g. isosceles, using ruler and compass and the measurement of the angles of the constructed triangle. Most of the children appeared to be able to use the protractor competently, except for a minority who, Jane later reported, were having difficulty.

At the start of the next programming session, Jane mentioned the difficulty she was still having teaching some of the children how to use the protractor. This difficulty was especially concerned with choice of the correct scale on the protractor. It turned out that her difficulty in teaching this skill was due to her own misunderstanding of why the protractor had two scales. She had not linked the LEFT and RIGHT rotations of the turtle, with the clockwise and anti-clockwise scales on the protractor.

I reminded her of the procedures LEFT and RIGHT and compared the rotations they produced with the protractor scales. She was encouraged to view angles, such as those in Figs. 6.12(a) and (b), as rotations by specifying,

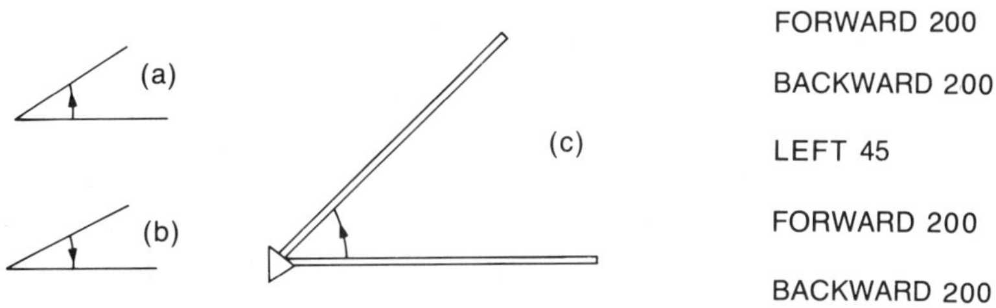

Fig. 6.12 — Angles as rotations.

in discussion, sequences of programming commands which would draw them, see Fig. 6.12(c). She did not actually program these commnds, she merely wrote them out on paper. Establishing this link between the school work with the protractors and Turtle Geometry proved beneficial as we shall see later in the description of a lesson observed subsequently. During the programming session she was asked whether she thought the earlier angle work had been helpful. She replied:

> 'Yes, well, it's just since I've been doing this that I realize, you know, I can sort of *visualize an angle*, you know if something is bigger than hundred and... ninety degrees. You know I can visualize how big its going to be. *But before it was dreadful because I managed to get them all mixed up*... I couldn't visualize them, no... unless I really, you know, thought about it hard for a long time.'

Ten days later, Jane was observed in school giving two girls remedial help in measuring angles with protractors. This lesson showed that Jane had applied some of her programming knowledge. Jane introduced the topic by asking the girls how they chose which protractor scale to use when measuring an angle. One replied, candidly:

'I just use the first one which comes to me.'

Jane had constructed devices consisting of two cardboard arms joined with a paper fastener at one end, see Fig. 6.13(a). She had made one for each

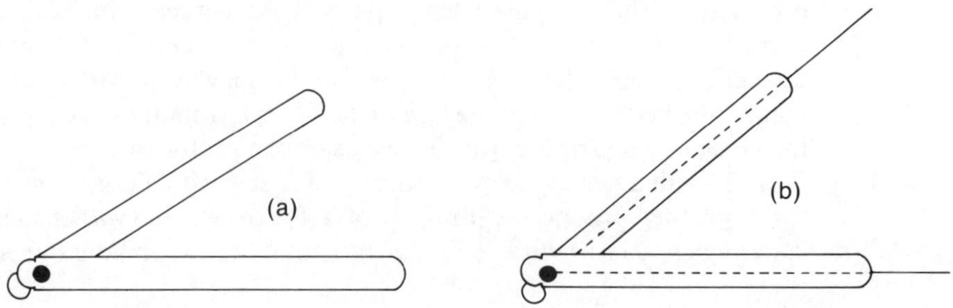

Fig. 6.13 — Teaching aid.

girl and one for herself to demonstrate with. She showed the girls how the device could be used to illustrate angles as rotations. This was done by first closing the two arms and placing them over one arm of the angle to be measured, with the paper fastener on the vertex. Then one arm was opened (i.e. rotated) until it lay along the other angle arm, see Fig. 6.13(b). Jane then tried to link this with the correct placement of the protractor and measuring from zero on the appropriate scale. Jane used the ambiguous terms 'left to right' and 'right to left' to describe clockwise and anti-clockwise rotations. Possibly these terms were derived from the turtle commands, LEFT and RIGHT. The girls attempted to match these terms to the inner and outer scales of the protractor. Unfortunately the girls only measured angles in *triangles*, and most of these triangles were orientated so that they had horizontal bases. This meant that the girls learnt rules of thumb to measure the slope angles of the triangles correctly. So one girl still had trouble measuring the angle opposite the base and tried to place the protractor as in Fig. 6.14. It was measurement of this 'top' angle of a triangle

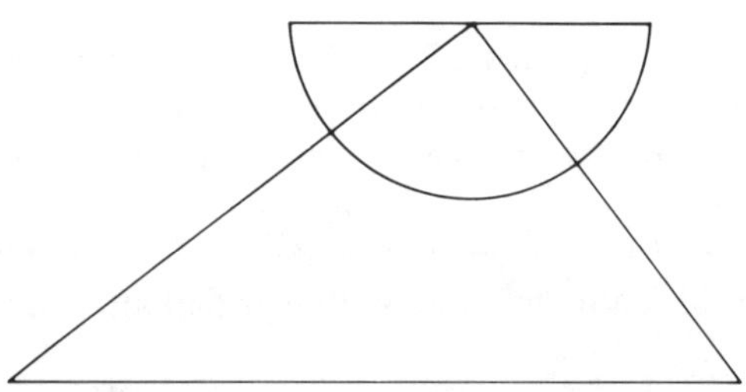

Fig. 6.14 — Top angle of triangle.

which Jane had reported she found difficult.

Later, Jane and Mary listened to the recording of this lesson together. I pointed out the ambiguity of the terms 'left to right' and 'right to left' and the restricted range of examples of angles which had been used. Mary suggested that Jane should have given more emphasis to the idea of 'starting at zero' on the protractor scale, which might have prevented the children using the wrong scale. Jane seemed to have the idea that only one scale of the protrator was the correct one to use for a given triangle angle. When I pointed out that either scale could be used for any angle (see Fig. 6.15) she

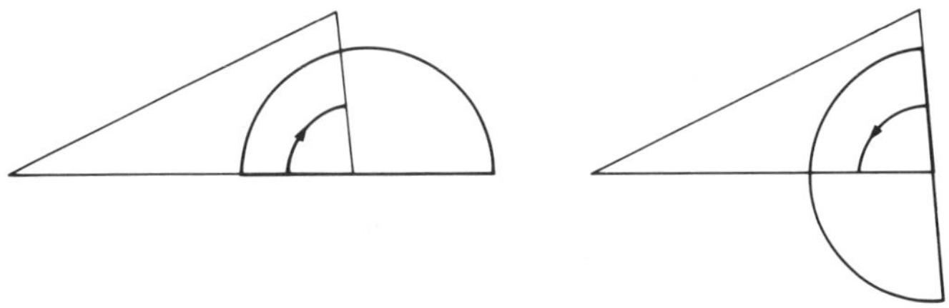

Fig. 6.15 — Using either protractor scale.

admitted, with some hesitation, that each way ought to give the same value for the angle.

The episode is revealing because it suggests that Jane had not linked the idea of angle as rotation, either clockwise or anti-clockwise, to angle measured with a protractor. This despite the extensive work she had carried out in Turtle Geometry. However, when this link was discussed, making free reference to her acquired programming experience, she was able to make some use of it to design a piece of apparatus to help her pupils. Again this emphasizes that the programming work should be accompanied by good teaching to help the students to make these links.

The next session also suggests that Jane had not linked together her school geometrical knowledge and Turtle Geometry. She was asked, as one question on a worksheet, to explain the relation between the angle rule for the triangle (that its angles sum to 180 degrees) and the fact that in a total circuit of a triangle the turtle turned exactly 360 degrees (the total turtle trip theorem). She was not able to do this. However, when filling in a chart of the angle properties of various regular polygons (another worksheet question), she noticed that the total interior angles of regular polygons formed a pattern. As the number of sides of the polygons was increased by one, so the total of the interior angles increased by 180 degrees. She started her chart with the square, rather than with the equilateral triangle, and so did not notice that the triangle also fitted the same pattern. Although she had spotted the pattern, she was not able to explain it. She ventured:

> 'Has it got to do with... the angles on a straight line?.. They are either one hundred and eighty or three sixty or...'

The problem was left unsolved for the moment. Jane then wrote a procedure which could draw any regular polygon, given its exterior angle and number of sides, and also compute and print out the total of its interior angles.

By the next session Jane had still not been able to work out why the interior angle total increased by 180 degrees. Mary, who had also been working on this independently and discovered the reason, tried to explain to her but failed. Finally I gave an explanation. Despite Jane's failure to solve the problem, something of mathematical benefit did occur. Jane spotted a pattern in the numbers in the chart and did try to find an explanation for this pattern.

Tessellating polygons

Jane spent five sessions drawing tessellations of different regular polygons. This task would probably have produced better mathematical benefits if it had not involved programming. These sessions show the way that problems could be tackled at the wrong level of representation because of the availability of the computer. In this session she drew a ring of pentagons (see Fig. 6.16) and then tried to fill the centre of the ring with more pentagons.

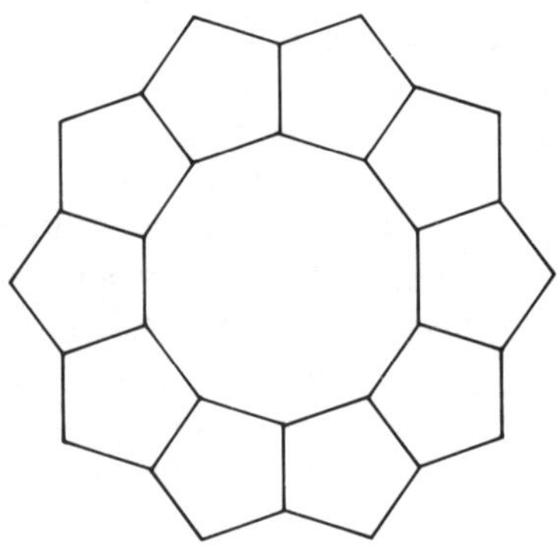

Fig. 6.16 — Ring of pentagons.

She found out that this could only be done if the pentagons overlapped, thus establishing that pentagons do not tessellate. In the succeeding four sessions she drew pictures of tessellating hexagons and octagons (which leave square spaces). However, most of her, extensive, programming work was devoted to incrementally building up the desired picture rather than to an analysis of why certain polygons would tessellate and others would not. She maintained she enjoyed this work which demanded a great deal of angle drawing. However, from the point of view of understanding tessellation not much was achieved for all the effort. Indeed some simple work with cut out paper polygons might well have established the same findings with a fraction of the time and cost.

The difficulty was that the programming problems of drawing pictures of tessellations was itself complex and intriguing. This diverted attention from the *mathematical* aspects of the situation. This may be seen as my failure to make the best use of all Jane's enthusiasm. I did not direct her strongly enough to search out, for example, why hexagons tessellate and pentagons do not. Perhaps if Jane had planned her pictures more thoroughly rather than incrementally building them, this point might have come out of her analysis.

ALGEBRA

In this section I describe Jane's work in various aspects of algebra. Some evidence is presentd which supports the claim that programming, in particular with Logo, can give insight into the key concepts of function and variable. It will be shown that the syntax of Logo provides an excellent illustration of 'function' and 'composition of functions'. An example is also given of the mathematics work Jane was tackling in the College of Education and how her programming work helped alleviate her difficulties, though not until I had explicitly pointed out the link between the College work and the programming.

Parsing mathematical expressions

The programming worksheets which Jane used in her first term of programming introduced such names as 'state', 'generalize', 'name' and 'value' in the context of specific programming activities, though without direct reference to mathematics. That term Jane also learnt about the parsing rules of Logo and about the use of parentheses to make the prefix commands of Logo more easily readable. She compared her own evaluation of how a command would be interpreted with the result computed by the computer. For example, by typing the following command:

```
PRINT SUM (PRODUCT 3 4) (DIFFERENCE 10 (SUM 3 2))
```

This prefix expression is equivalent to the more conventional infix expression given below:

$$(3*4)+[10-(3+2)]$$

Use of parentheses was not mandatory in Logo commands, but it was recommended. The quite explicit distinction between the prefix formulation of an arithmetic expression in Logo and the more conventional infix notation is a valuable asset because it forces the issues of the binding, scope and parsing of arithmetic operations into the open. It was for this reason that the infix arithmetic operations were removed from our final implementation of Logo.

The intention of these exercises was to illustrate the idea of 'parsing' a mathematical expression, by asking Jane to translate backwards and forwards between two notational systems. Further examples of this kind were undertaken by Jane when she returned to programming after her first teaching practice in the summer. In the second questionnaire administered that term, Jane mentioned this work in her answer to the question about the value of the programming work in her future teaching. She mentioned the work with angles (see section 6.2) and also:

> '...use of brackets in some procedures to make an inner procedure to make the answer easier to find—I find that using the computer to do them much more enlightening than just being told about it.'

Here use of the word 'inner' suggests that she has grasped one of the essential uses of parentheses.

Variables and functions

In her study of variables, Jane wrote a number of procedures taking arguments. Jane extended a given rectangle procedure to compute and print out its area and perimeter. She had some difficulty in forming syntactically correct commands to compute the area and perimeter, although she was familiar with the algorithm for their calculation. The main difficulty was in her use of the *name* of the variable instead of its *value*. Eventually, with some help from me she sorted this out. She chose to use parentheses in the commands which computed area and perimeter.

After further practice in using argument names and values, Jane wrote a number of procedures which acted like functions. These took a single argument and computed a single result, which if necessary could become the value of the argument to another procedure. One such function-like procedure was to square numbers (i.e. 3→9, 5→25). In Logo such a procedure, named SQUARE would be used as follows:

```
W: PRINT SQUARE 5
 25
```

Here the prefix notation of Logo and Jane's experience in parsing commands made it easy for her to understand both flow of control and result passing in the composition of SQUARE procedures below:

```
PRINT SQUARE SQUARE SQUARE SQUARE 2
```

though she was a little surprised at the number printed, 65536.

Author: What did you work it out to be?
Jane: I thought maybe about thirty-two.
Author: Yes right that... yes, yes... Do you know why it isn't thirty-two? Why?
Jane: Because you keep, you keep having to square the number.
Author: Right.
Jane: And I kept multiplying up by two.

I pointed out that this was a common mistake, by explaining how I had jumped to the same conclusion when first encountering this exercise! Jane continued:

> 'I find I always make mistakes with squares because, huh, we had to do the slide rule in one of our maths things and when it said square, I just multiplied by two, forgetting that you are actually having to multiply by... by itself.. I always... the minute I see square, I always think its because, I think of the little two at the side. Nine squared, you can write it as nine with a little two. I always think of multiplying by, by two instead it means multiplied by itself.'

This incident had the value of underlining the difference between squaring and doubling because of the disparity between Jane's hypothesis of what the computer would print and what actually was printed. They are easily confused because of their similarity of notation:

5^2 and 5×2

Logo's rather long-winded formulation of the procedure for squaring a number is quite different from the procedure which will double a number. Compare the two procedures below, which have been given the appropriate names:

```
DEFINE "SQUARE "NUMBER
10 RESULT PRODUCT VALUE "NUMBER VALUE "NUMBER
END

DEFINE "DOUBLE "NUMBER
10 RESULT PRODUCT VALUE "NUMBER 2
END
```

Another value of programming in the teaching of functions is that, in a prefix language like Logo, it is quite straightforward and natural to compose functions, as in the multiple use of SQUARE, above. Also the passing of results in the command corresponds with that function notation convention which interprets $fg(x)$ as applying function g to x, and then applying function f to the result. This point, misunderstood by both Jane and Mary, will be returned to later.

Equation solving

After the summer vacation, Jane was again on teaching practice and this time also attending programming sessions. At her first session that term she explained a difficulty which had arisen on her first morning of teaching practice. She had been asked to help children do problems in equation extraction and solution. She found it difficult to extract a single variable equation from a problem stated in English and difficult to solve such an equation once it was extracted.

In attempting to help her, use was made of her previous programming experience of names, values and parentheses. I pointed out that solving a story problem consisted of solving two sub-problems. The first was to successively transform the equation until it reached a form which could be solved by inspection. It was explained that the 'letters' used in the equation in the book, she had brought from the school, were 'names', similar to argument names in programming. It was also pointed out that the process of solving such an equation was that of finding the appropriate value to associate with that name. Jane was shown how parentheses could be used to mark successive stages in the extraction of an equation from the text and then tried the following problem herself.

> 'I am thinking of a number. When 5 is subtracted from six times the number, the answer is 37. What is the number?'

With a little prompting, Jane wrote out the following sequence of expressions in which the variable was gradually imbedded more deeply by means of parentheses. It was suggested she use the name 'number' for the variable. The following diagram shows what Jane wrote together with the appropriate phrase from the problem text.

Equation extraction	*Problem text*
number	I am thinking of a number
(number*6)	six times the number
((number*6)−5)	when five is subtracted from
((number*6)−5)=37	the answer is 37

Solving the equation was shown to be a reverse process of evaluating the parentheses:

((number * 6) −5)=37
(number * 6)=42
(7 * 6)=42

Jane was asked to express the equation more conventionally, and correctly wrote:

$$6x - 5 = 37$$

This work used a variety of terms derived from programming and was conducted 'off the cuff' in response to Jane's enquiry.

Symmetry transformations

Jane's work with tessellations was an example where the programming task was intriguing but mathematically unproductive. In contrast, the following work on transformations was ideally suited to exploration through (Logo) programming. The distinction between 'state' and 'transformations' emerged clearly. The programming details were reduced by providing the students with new primitives for this domain. Their task was to run combinations of these primitives and observe their effect.

After teaching practice Jane was again attending lectures in the College of Education. Both Jane and Mary complained that they were finding some of their mathematics course hard to follow. They had both chosen to take mathematics as one of their principal areas of study because they knew that they needed extra help. Others taking the same course had been more successful in mathematics and it appeared that some topics chosen in the lectures, while suitable for the majority, were causing Mary and Jane some difficulty. In particular they complained about some work based on groups, including the group of symmetry transformations of the rectangle. Jane felt worried both about the overall objectives of this work and about how she was to do the questions on a worksheet supplied by the College.

> 'Well I'm just not sure exactly what we are supposed to be doing. Why, I mean... what is this that we are trying to do? I don't see it at all... and I'm not sure how to do it anyway. I mean I have done bits of it, but I don't see why I am doing it. I think you are following a formula but I don't see why I am doing it.'

Her comment 'following a formula but I don't see why' is like many of her remarks, which reflect her knowledge of mathematics as rules without understanding the reasons for those rules. Two weeks later I had prepared a worksheet and written procedures based very closely on the examples and notation in the worksheet from the College. The four symmetry transformations of the rectangle were modelled in four procedures which transformed and drew a rectangle on the display screen. One corner of the rectangle was marked with a spot so that the effect of a transformation could be detected. The transformations and their names and effects are given in Fig. 6.17.

I Identity transformation
X Turn over about horizontal axis of symmetry
Y Turn over about vertical axis of symmetry
Z Rotate in own plane through 180 degrees about centre

I wrote these procedures because they were beyond Jane's ability to construct and because it was felt that *writing* procedures to illustrate a transformation by drawing was not necessarily the best way for Jane to understand the properties of the transformation. The procedures were written in Logo, in such a way as to allow the composition of the transformations, e.g. a reflection followed by a rotation. The worksheet from the

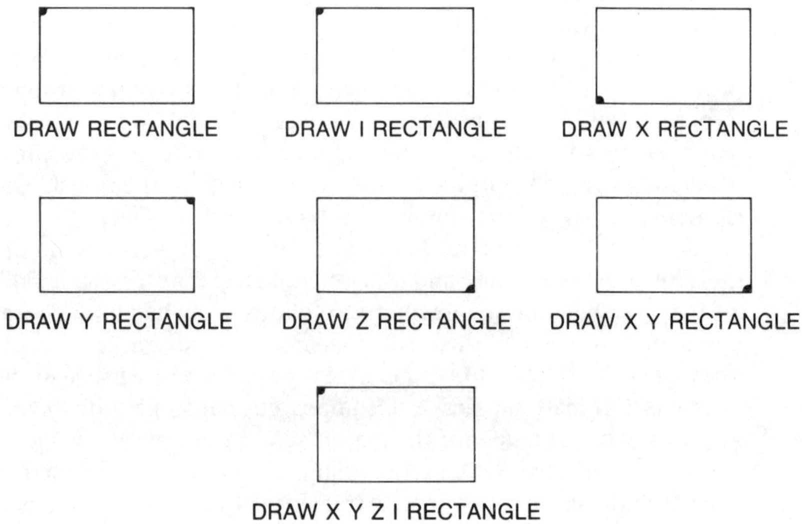

Fig. 6.17 Symmetry transformations.

College of Education invited the student to fill in a group table for this set of transformations together with the binary operation 'followed by', see Fig. 6.18. They were also given a graph whose four nodes corresponded to four

followed by	I	X	Y	Z
I	I	X	Y	Z
X	X	I	Z	Y
Y	Y	Z	I	X
Z	Z	Y	X	I

Fig. 6.18 — Group table.

orientations of the rectangle and were asked to label the arcs between the nodes with appropriate transformations, see Fig. 6.19.

One of Jane's original difficulties had been that she had not clearly distinguished between a transformation of the rectangle and the state of the rectangle before and after a transformation was applied. She had labelled the nodes as well as the arcs with the names of transformations. Another of her difficulties concerned composition of transformations. The College of Education worksheet explained that in its notation ZX meant first do transformation X and then do Z to the result. What Jane had not fully grasped was that Z was to be applied to the rectangle's state produced by first doing X. She thought that one chose a state, did X, then chose the initial

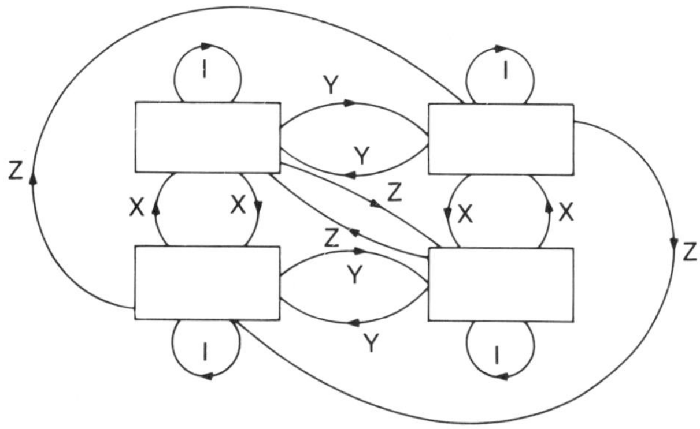

Fig. 6.19 — Graph of transformations.

state again and Z. She also explained about the difficulty she had earlier, trying to explain to her College lecturer what she did not understand:

> 'Well, you see the bit that I tried to explain to Mr. B. [in the College of Education] that I didn't understand, but he didn't understand what I was trying to explain... was that when you do one operation followed by another, how isn't it that *you just get the, the last operation as your answer*? For example if you do X followed by Y, why isn't it that its just Y that's the answer?

In her system every composition would be equivalent to the transformation applied last of all. That is:

$$ZX=Z$$
$$XYZ=X$$

This latter difficulty was especially apparent when she had to make up entries for the group table, see Fig. 6.18. Jane reported her difficulty as follows, in relation to a similar piece of College work on the symmetries of an equilateral triangle:

> '... I could see when we had a triangle and were twisting it around in different ways. But when you were making up the tables, that's when I really get lost. I don't see how, you know, he [the College lecturer] gets all these letters... I really get lost with that.'

She went on to explain that she understood how each transformation would allow one to pick up the shape and replace it in an identically shaped frame. She also realized that different shapes would have different sets of such transformations, but she added:

> '... but its when you put them in a table, that's when I find it hard... you know, one following another.'

During this session and the next Jane worked on the worksheet which provided a visual illustration, using the procedures, for the work set by the College.

The programming work was valuable because it distinguished between the state of the rectangle (as seen on the screen) and the transformation (typed in as part of a command). The prefix syntax of Logo exactly matched the notation adopted by the College. Jane then knew about result passing (see the SQUARE example earlier) and I was able to point out the similarity.

However, Jane had not perceived this similarity for herself. In this way it was like the protractor/angle example seen in section 6.1. Jane had covered all the programming prerequisites but had not made the link between the College work and the programming work. Once the link was pointed out, however, she was able to understand it.

In the course of running the procedures written by me, Jane spotted a bug. This was that transformation Y did not leave the rectangle in the correct state. Jane pointed out that Y was having the wrong effect, and I fixed the bug. This suggests that although Jane was only running my procedures, she was observing their effects carefully and matching these effects to the worksheet text.

In the second session the difficulty about composition of transformations was attended to. Jane tried out a variety of commands which involved two transformations to see what single transformation could have produced the same effect, e.g.:

```
DRAW X X RECTANGLE
```

This had the same effect as

```
DRAW I RECTANGLE
```

I explained the composition of transformations by asking Jane to consider the overall effect of two transformations, 'painting the rectangle blue' and 'turning the rectangle upside-down'. After the discussion Jane explained that she now knew how composition worked.

> 'Oh I see it's the two together.'
> 'So its the result of both of them together.'

She was then able to fill in the group table, see Fig. 6.18, and to solve questions involving finding a single transformation equivalent to a composition of two or more transformations. For example, she found that there was no transformation P which satisfied the equation below:

PX=P

However, she still had labelled her graph incorrectly by marking both arcs and nodes with the transformation names I, X, Y and Z. Her difficulty had been made worse because the arcs for the identity transformations had been omitted from the diagram given to her. The difficulty was further aggravated because there was a preferred initial state for the rectangle, produced by the following command:

DRAW RECTANGLE

This command specified no transformation and merely drew the rectangle. Jane labelled the other rectangle states on the graph with the names of the transformations, which would produce them from the initial state. What she had not fully grasped was that the particular initial state of the rectangle was immaterial to the relations between the transformations that held for all possible rectangle states. That is, the statement:

XY=Z

was true whatever the initial state of the rectangle. I tried to explain this point but Jane took little active part in the conversation, merely saying 'uh-huh' at intervals, so I was unable to gauge how far she had understood. In this respect, Jane behaved quite differently from Mary, who would not let me explain 'at' her but would take an active part in the conversation by demanding further information or by explaining back to me what I had just said. At the end of this explanation Jane reported that:

'I feel a bit more confident with these ones.'

But she said she was still a little worried about dealing with other shapes. She mentioned the 'Isle of Man' symbol in particular.

NUMBER

Jane was worried because she knew a number of mathematical rules but did not know why these rules worked. A typical example was the rule about dividing fractions. This section describes how Jane used procedures, provided for her, which gave a visual illustration for such rules. An example is also given of a project with fractions which was abandoned and which shows how it is possible for the programming aspects of a problem to overwhelm the mathematical aspects of that problem. Part of the success of using pre-written procedures was that they allowed Jane to explore mathematics difficulties at the right level of representation. In the abandoned fraction project she spent her time worrying about drawing fraction diagrams rather than worrying about fractions.

Drawing fraction diagrams

Jane attempted two worksheets concerned with fractions. The first worksheet attempted to show how fractions were a new type of number built on the natural numbers. It also attempted to show how fractions could be considered as a 'double-operation' of division and combination. This latter was to be illustrated by the student defining a general procedure, named FRACTION, to draw fractional parts of a disc, see Fig. 6.20. This procedure

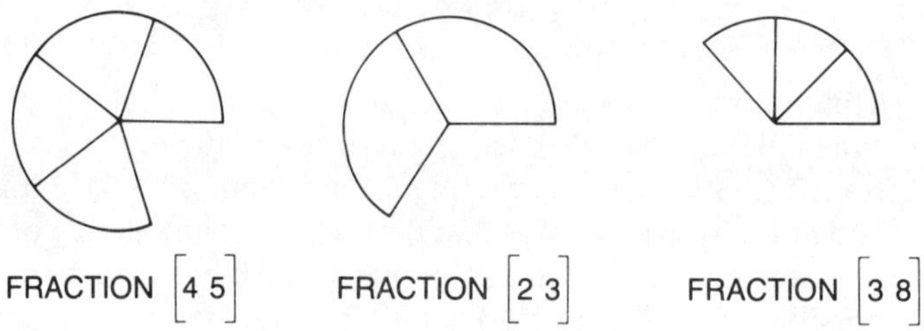

Fig. 6.20 — Parts of a disc.

took a single argument whose value was a list of two elements representing the numerator and denominator of the fraction respectively. Students were intended to break down the problem of drawing such a diagram into sub-problems. One sub-problem was to draw a single 'slice' of the disc (illustrating the division aspect of a fraction) and the other was to combine a number of these slices together (illustrating the multiplicative aspect of a fraction).

When Jane attempted the problem she subdivided the problem in a quite different way. She tried to draw the 'spokes' of the chart separately from the arc of the circle, as in Fig. 6.21. Although this was not what I expected, or

Fig. 6.21 — Problem decomposition.

wanted, I decided to let Jane pursue it. The procedure she started to define took the radius of the circle and angle between the spokes as two arguments. These arguments were expressed in units of length and in degrees and were not the dimensionless quantities suggested in the worksheet. By the end of the session, Jane had still not fully debugged her procedure which would work only for a particular angle argument value. This was the last session of the term before the Christmas vacation. About a month later, Jane returned to the problem and went some way further in debugging the procedure. This session ended suddenly with a power cut, the whole building was plunged into darkness and the system crashed.

The next session lasted only fifty minutes before there was another system failure. In that time Jane got little further with the problem than in the last session. In the end the worksheet was abandoned and the next session Jane went on to the work on transformations of the rectangle, described in the last section.

Despite the unfortunate sequence of crashes, this particular project was unsuccesful because Jane had planned a different programming solution than that expected. She had also become immersed in the problems of producing one of the pictures given as an example in the worksheet, as in the tessellation project described in section 6.1. Both of these factors diverted her attention away from the proposed objective of understanding fractions.

Putting meaning into fraction rules

Much later Jane worked on fractions again, this time running procedures written by me. These procedures were intended to illuminate a meaning for multiplication and division of fractions which Jane had explained she did not understand:

Jane: Why is it, when you div... why is it when you divide.
Author: You turn it upside-down and multiply.
Jane: I don't understand why you do that because... we did the division... we did it last term at [College] and the tutor, I had, you know, we all asked why and he said it was obvious why, and never really explained.
Author: It's not the least bit obvious.

At the time I made some general comments about addition and subtraction of fractions being easier to illustrate than multiplication and division. It seems that my own attempt to explain the rule for dividing fractions, in a seminar in the first term, had not been remembered. Eventually a worksheet and a set of procedures were written by me which addressed this issue and which provided an illustration for multiplication and division of fractions.

Jane was given a procedure to run which drew a simple house on the display screen. The procedure then requested her to type in two numbers representing the numerator and denominator of a fraction. This fraction was then used by the procedure to draw a second house of similar shape but changed in size according to the value of the fraction. Thus an entry of '3 4' would produce a second house three-quarters the size of the first, while a subsequent entry of '2 1' would produce a third house twice as large as the second, see Fig. 6.22. This process could be continued indefinitely, so long as the student did not enter a fraction such that the resultant house was too large for the screen.

The essential feature of this procedure was to provide a ratio or exchange interpretation of fractions which could be used to illustrate multiplicaton by composition of exchanges and division as the inverse of multiplication.

Jane's activity consisted of running this procedure in response to questions posed in a worksheet and of discussing the meaning of the questions and the visual effects with the author. One reason for providing such a procedure rather than suggesting Jane wrote one herself was the long gap since she had last programmed. A second reason was the failure of her last fraction project where programming considerations had overwhelmed the mathematics.

After finding out how to run the procedure and the kinds of effect it could produce, Jane attempted to answer questions on the worksheet:

> Find out what fractions leave the size of the house unchanged (the identity).
> Find out what fractions have the same effect as 3 4, i.e. 3/4.
> Find out what fraction exchange undoes the effect of 3 4 (the inverse).

Jane already knew something about fractions and had ideas on what the answers to some of these questions would be. The session did not consist of her 'discovering' entirely new fraction concepts but of her confirming, clarifying and putting meaning into the various fraction 'jingles' which she knew ('turn upside-down and multiply').

She saw that fractions such as 1/1, 6/6 did not change the size of the house and described how an infinite set of such fractions could be generated. She had some trouble, initially, in deciding on an equivalent for 3/4. Part of the trouble seemed to lie in her lack of understanding of what was required. She suggested that three-quarters multiplied by itself might be equivalent but then rejected it because it produced a smaller house. She then discovered the equivalence:

$$3/4*1/1=3/4$$

This was like earlier work with symmetry transformations where the task had been to find a single transformation equivalent to a given *pair* of transformations. Now Jane appeared to be looking for a pair of fractions, to be multiplied together, which were equivalent to the single given fraction, 3/4. At this stage I explained what was required, using hand-drawn schematic diagrams based on the display pictures, see Fig. 6.23(a).

Using the diagram I showed how one 'path' from house A to house B was via the exchange 3/4. I asked her what other 'paths' could be found which linked the two houses. She was then left to solve the problem, which she did having tried 6/8 and 9/12 using the procedure. Again she gave a rule for

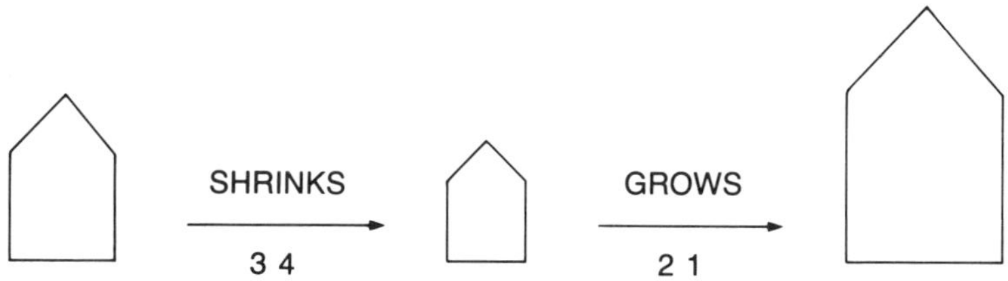

Fig. 6.22 — Multiplication of fractions.

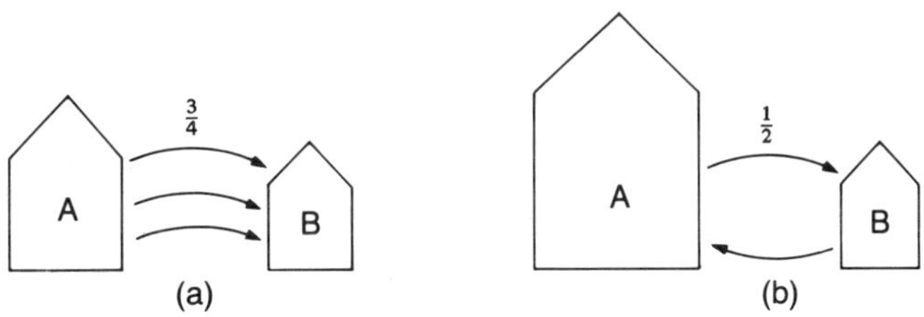

Fig. 6.23 — Equivalent and inverse fractions.

generating an infinite set of equivalents for 3/4, and related this rule to that for finding equivalents for 1/1, which she said was 'just the same idea'.

I asked her whether she knew the phrase 'family of equivalent fractions' or 'set of equivalent fractions'.

Jane: Yes, we have done that. *I know what it means now, because I was never really clear exactly before... exactly what that meant.* But it's just the same fraction multiplied by the... the same as a fraction multiplied by the same number again and again.
Author: Em, yes, you mean above and below.
Jane: That's right.

Her explanation of equivalence is a little ambiguous, and is expressed as a rule for generating equivalent fractions rather than in terms of the visual exchange interpretation. Nevertheless she now seemed to understand what equivalence meant.

When she was looking for the inverse of 3/4, I again used the schematic diagrams to suggest that the inverse of a fraction was the 'path' *back* from the second house to the first, see Fig. 6.23(b). This time she suggested that 3/4 might be its own inverse but rejected it without use of the procedure because she realized that

$$3/4*3/4 < 3/4$$

She then said:

So would you have to *change the fraction round*.

I was neutral about this suggestion and left her to work with the procedure. She tried the sequences schematically indicated in Figs. 6.24(a) and (b).

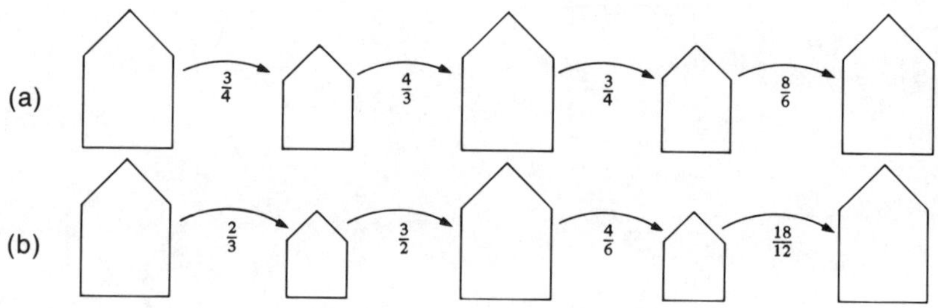

Fig. 6.24 — Fraction sequences.

When I returned she explained that:

'You just have to turn the fraction round.'

She also gave some examples. But she also added, unprompted that '*any equivalent fraction*' could be used.

Both her exploration with the procedure and her answer suggest that she had formed a clear idea of equivalent fractions and was able to see, for example, that 18/12 was the inverse 2/34. When asked why she had tried 4/3 as the inverse of 3/4 she replied:

'If you divide a fraction by itself turned round you get back to one'

This was incorrect since she should have said 'multiply' for 'divide'. However, it was passed over at the time, since Jane did link together ideas of 'division', 'inverse' and 'getting back to the beginning'.

The last question on the worksheet was for her to find a single fraction equivalent to the pair 3/4∗1/2. Again the question was described schematically as in Fig. 6.23. She knew that the answer should be 3/8 and confirmed it using the procedure. She did this by showing that 8/3 was the *inverse* of 3/4∗1/2. That is, it exactly undid the combined effect of 3/4 and 1/2.

This session ended at this point. Jane said that she had enjoyed the session. Her demeanour throughout had been rather different from previous sessions. This time she took a much more active part in conversations with me, preferring explanations and enlarging on what I had said. In earlier sessions she had tended to listen politely to an explanation but not try to re-explain back to me.

The next session built on what had been already achieved to illustrate a meaning for fraction division using the same schematic diagrams and by running the same procedure. In the course of answering the questions on the worksheet, Jane was able to evaluate correctly the expression:

$$(1/2)^2 \text{ and } (1/2)_3$$

She mentioned that she no longer muddled up squaring and doubling as we saw earlier. The worksheet used the term 'commutative' which she could not remember the meaning of, except that it had something to do with addition. When she was reminded of the meaning she clearly understood that fraction multiplication was commutative.

To illustrate fraction division, the worksheet suggested that a division should be restated as a multiplication, as below:

$$8/2=X \Rightarrow 2*X=8$$
$$(2/3)/(1/4)=X \Rightarrow (1/4)*X=(2/3)$$

In each case the division is solved if the variable in the multiplication can be found. Later she mentioned that she had to check such restatements for the case of fractions by matching them against an example using natural numbers. A similar strategy was also observed in her mathematics test answer to a formula rearrangement. The multiplications derived from these rearrangements of the problem were to be illustrated using the same procedure as in the last session, see Fig. 6.25. Jane spent some time trying to

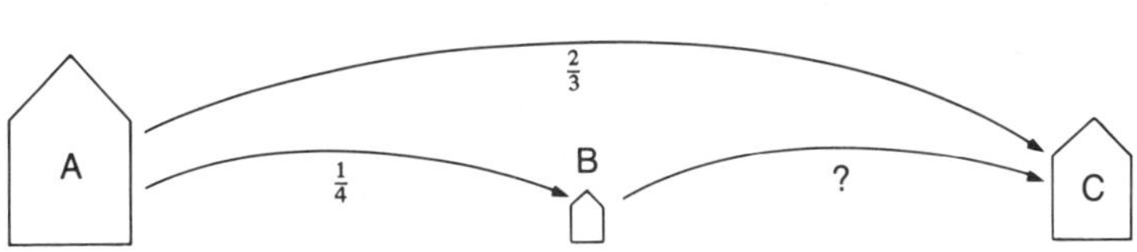

Fig. 6.25 — Division of fractions.

find the value of the 'path' marked with a '?' in the diagram, using the procedure. Finally I re-explained the problem in terms of the schematic diagram. That is to find the missing fraction between house B and C, one could take the alternative but equivalent path from B to A (using the inverse) and then from A to C. Suddenly she understood:

'Oh I see, so its just four over one multiplied by two over three.'

What she had not appreciated earlier was that the exchange B→C was equivalent to the exchange B→A→C, see Fig. 6.26.

Jane then mentioned the problem of teaching children fractions and worried that this presentation of fraction operations was not applicable in the classroom.

> 'The thing is could you explain that to children?'
> 'How often... I mean, *I have never seen a teacher do it* and it's a case of there's a rule and the children have to follow it because I don't think they ever do drawings where they have to change the shape.'

Jane's comments illustrate a difficulty of teaching mathematics to student teachers. She was worried that she could not apply the approach used here, in her own classroom. Her concern was not whether she understood this fraction concept, but whether she had a technique available to teach the children.

I suggested ways in which the presentation might be adapted for the classroom and reiterated my belief that one had to try to teach for understanding. Otherwise one would only produce more students, like Jane, who knew rules but did not know reasons for those rules.

A little while later she explained on the same theme, in a way which suggests that she had benefited from the fraction work:

> 'I know myself that if I had to... probably if I hadn't done this and some child had asked me... I wouldn't 've... aha, I would say *it's just a rule*. And I suppose a lot of teachers maybe just have to say that... it's just a rule.'

Coordinates and vectors

This section describes Jane's work with coordinates. It shows how work with coordinates, taking negative integer values, was incorporated naturally into her programming work because it solved the problem of accurately positioning a picture on the screen. In this case mathematics was seen as useful in solving a problem, as it had been in the case of drawing an 'arrow' (see section 6.1).

During the summer vacation, Jane was given a worksheet on coordinates to help her position pictures on the display screen more easily. She found the worksheet hard to follow at first since she had very little prior knowledge of this topic. I spent some time explaining coordinates both in Turtle Geometry and in other contexts, e.g. map reading. Jane was still anxious and decided to postpone work on this topic until the next session. She found new terms such as 'axis' confusing.

The next session, Jane successfully defined a procedure of two arguments, named GRID. This procedure centred the turtle on the display screen and would move it anywhere within the top right-hand quadrant of the screen by using the two arguments as x and y coordinates, see Fig. 6.27. Although Jane understood that this procedure could be used to position the turtle anywhere in the shaded portion of the screen, she suggested that different procedures would have to be defined to move it to the unshaded

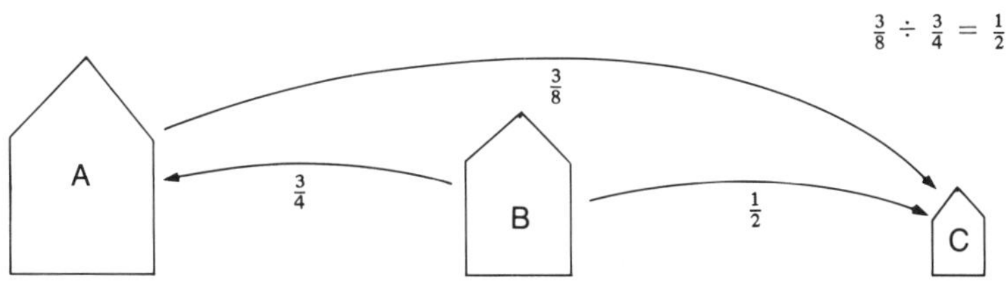

Fig. 6.26 — Equivalent paths.

```
DEFINE "GRID "ALONG "UP
10 CENTRE
20 LIFT
30 FORWARD VALUE "ALONG
40 LEFT 90
50 FORWARD VALUE "UP
60 DROP
END
```

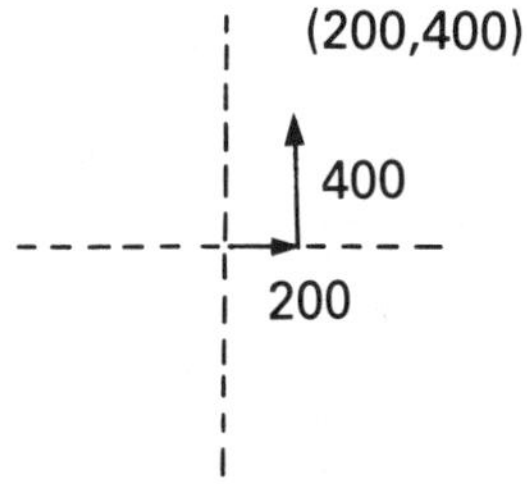

Fig. 6.27 — Coordinate plotting procedure.

parts. She did not realize that, merely by giving her procedure negative argument values, the turtle could be moved anywhere on the screen. I then introduced the idea of negative integers and suggested that she find out the effect of giving negative input to the drawing procedures FORWARD, BACKWARD, LEFT and RIGHT. When I returned, she reported:

> 'Well a minus seems to make it go in the opposite direction.'

But she still did not see the relevance of this to the coordinate problem and asked:

> 'So why do you use minus numbers then? Why can't you just say backward ten [instead of forward minus ten]?'

I did not answer the question but left her to think about it for herself. She then tried out her coordinate procedure with a variety of negative and positive integer arguments and found out that she could in fact position the turtle anywhere on the screen.

In discussion with me she was able to explain the relation between the sign of the argument values for her procedure and the quadrant into which the turtle would be sent, as in Fig. 6.28. She also explained that when working on this problem at home, before the session, she had decided to

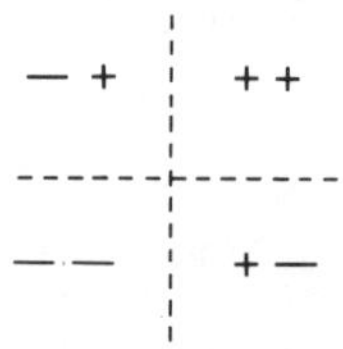

Fig. 6.28 — Signs of coordinates.

place the origin of her coordiante system in the bottom left-hand corner of the screen, rather than in the centre. This would have obviated the need for negative integers. This was a valid solution to the problem, but I wanted her to see how negative integers could be useful. This proved to be the case because she used her procedure GRID in subsequent sessions, with negative arguments, to position the turtle (for example when drawing spirals).

In the spring term 1977, Jane had to teach single lessons on specific topics to small groups of children as part of her College of Education mathematics course. For one of these lessons she had been directed to teach vectors and was rather worried about it. She described a game on a grid in which children were to use two element vectors to make moves. I discussed this with her and related it back to the coordinate work she had done before, including the work with negative integers. It was suggested that the children could write 'procedures' consisting of sequences of vectors which could then be plotted to produce outline shapes as in Fig. 6.29. The similarity of this to Logo

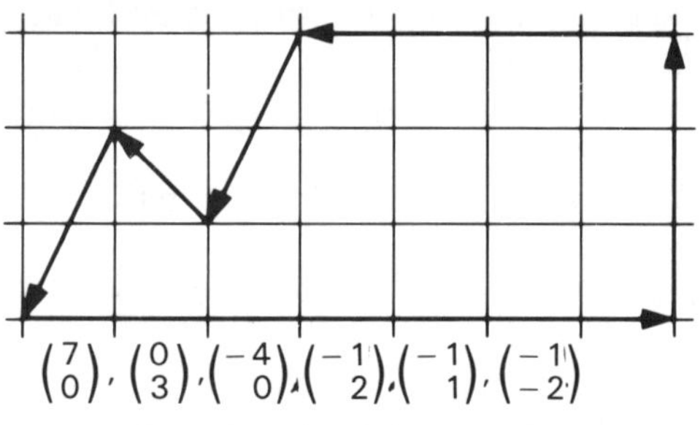

Fig. 6.29 — Sequence of vectors.

drawing was also brought out. It was also suggested that the turtle state of [position, heading and pen inclination] could be considered, by Jane, as a three-dimensional vector. However, the discussion rather muddled up the concept of coordinate with that of vector. The distinction between vectors and coordinates was further confused by the discussion of integers which were illustrated as directed numbers. These were specifically explained as one dimensional vectors showing '+' and '−' as signs indicating direction.

Some work was undertaken with vectors by considering turtle movements as relative vectors. This was a mistake since turtle movements did not behave like vectors, e.g. addition of movements was not commutative. This meant that Jane was working with an incorrect set of primitives. The failure of this work was similar to that of the fraction pie-chart drawing, considered earlier. In each case the task of the student was to draw pictures representing the structure in question. But commands to *draw* a representation of a structure (e.g. a vector) are not the same as commands to manipulate that structure. In the same way, commands to manipulate symbols representing fractions may reveal little about fractions. Jane was asked to draw the pair of vectors AB and BC, in Fig. 6.30, using Logo. She was then asked to find

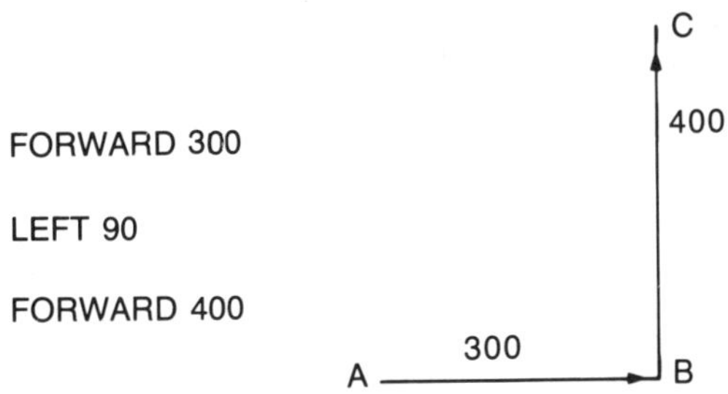

Fig. 6.30 — Drawing vectors.

their sum by seeking for the single command which would produce a line corresponding to AC. She soon discovered that the magnitude of AC was not equal to the sum of the individual magnitudes of AB and BC. She did not automatically calculate magnitude using Pythagoras' theorem but tried a convenient first order approximation that slope angle was 45 degrees and that the magnitude was given by $|AC|=|AB|+|BC|$. By a process of successive refinement she found the best answer on her sixth attempt, see Fig. 6.31.

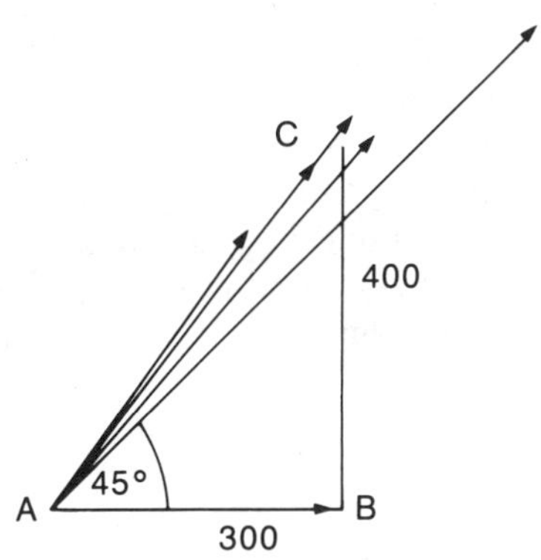

Fig. 6.31 — Finding the sum by successive refinement.

Jane spent much time on another question which asked her to show that addition of vectors was commutative, using Logo drawing. She tried to do this by driving the turtle back from C to A, using the same figure as the last question,above. She was plotting the opposite vectors CB and BA. She failed to draw these two vectors because she could not get the heading of the turtle correct at point C, see Fig. 6.30, to draw the first vector CB. Instead she produced Fig. 6.32. In this figure the numbers 1, 2 and 3 indicate three

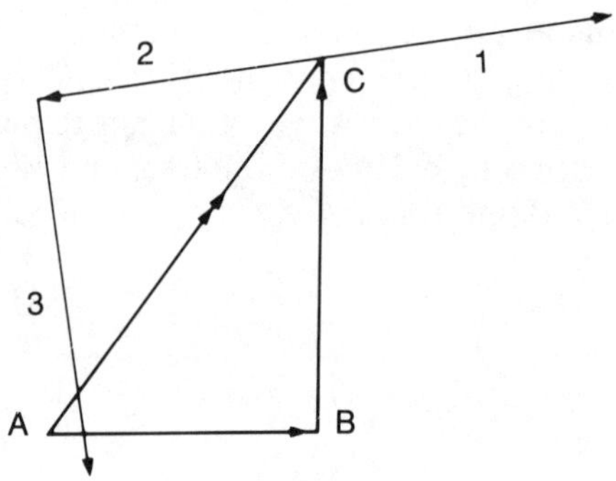

Fig. 6.32 — Three attempts to draw a vector.

attempts to draw the vector. This work provides an example of the inappropriate use of Turtle Geometry primitives where vector primitives should have been provided.

Modelling integers

The integer work shows how the standard drawing primitives FORWARD and BACKWARD were successfully used to model integers and integer operations via the idea of directed numbers. It also shows how Jane was able to solve problems with this modelling system and also how she was able to extend the scope of schematic diagrams based on the system.

The model for a positive integer was a call on the procedure FORWARD, and for a negative integer a call on the procedure BACKWARD. Each new sequence of calls started at the centre of the graph-plotter paper (graph-plotters were used for this work rather than the display, because it enabled Jane and Mary to sit next to each other, each with a graph-plotter). The model for addition of such integers was to follow one movement by another, starting the second movement from where the first finished, see Fig. 6.33. In the figure the movements are arrowed and distinguished

(+ 40) + (—50) = (—10)

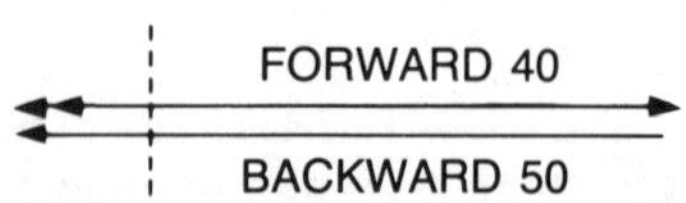

Fig. 6.33 — Model for integers.

vertically. When Jane (and Mary) drew them they were plain lines, sometimes passing over each other, it the paper was not moved vertically.

At first Jane misunderstood the model. She thought that, in the model, the sum of the movements was that movement which would return the turtle pen to the starting point; see Fig. 6.34. She did not see that it had to be that

(+ 40) + (—50) = (+ 10)

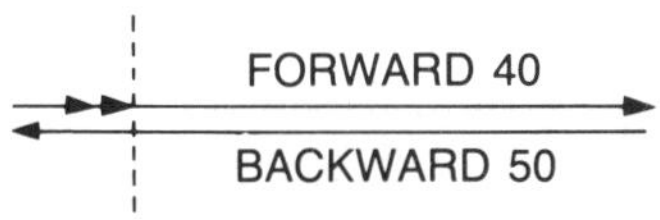

Fig. 6.34 — Incorrect sum of two integers.

single movement which would have the same overall effect as the component movements taken together.

I did not realize initially that Jane misunderstood the model. Early questions of hers should have alerted me:

Jane: Do you still call it minus *even though it's not really minus*?
Author: Now what do you mean, it's not really minus?
Jane: Well it's not really addition and subtraction but well that's... that answer there, that vector.

Her problem here appeared to be the common confusion between addition and subtraction as operations and plus and minus signed integers.

Jane and Mary were sitting side by side working individually on the same problems on two graph-plotters. Jane looked over to Mary and saw that she was getting different answers to the worksheet questions. Jane then corrected Mary and the following conversation brought out the nature of *Jane's* mistake:

Jane: that should be plus fifty.
Mary: No, whatever is on your left-hand-side is minus, isn't it.
Jane: But you are having to go forward to get back to the centre again.
Mary: So, if you want to go back to the centre, you have to tell it plus fifty.
Jane: So it's not minus fifty.
Mary: It is. I mean, how, I mean, in which direction has your plotter moved? That is what you are trying to find out.

Mary had concisely stated what needed to be done to find the sum: that is, see how much and in what direction the plotter pen had moved as a result of the given sequence of commands. This was contrasted with Jane's view that the sum of the movements was that movement needed to 'undo' the effect of all the commands. With some further explanation from me, Jane understood Mary's point.

The next exercise asked the students to write 'banking' procedures named DEPOSIT and WITHDRAW which were intended to model 'movements' of money into (plus) and out of (minus) a bank account. I had intended the students to increment and decrement a variable. But Jane solved the problem by using the drawing model above. Her procedure for depositing money moved the plotter pen forward, to the right of the paper, and her withdrawal procedure moved the plotter pen backward, to the left of the paper. The balance was shown by the pen's current position. To the left of the centre the account was in debit and to the right it was in credit. Jane needed a little help in debugging her procedures because of programming errors. The finished procedures are shown below and part of the sequence of Jane's use of them is given in Fig. 6.35.

```
DEFINE "DEPOSIT "ACCOUNT
10 FORWARD VALUE "ACCOUNT
20 PRINT [THE BALANCE IS]
30 PRINT XCOR
END

DEFINE "WITHDRAW "MONEY
10 BACKWARD VALUE "MONEY
20 PRINT [THE BALANCE IS]
30 PRINT XCOR
END
```

```
W: DEPOSIT 20
  [THE BALANCE IS]
  100
W: DEPOSIT 30
  [THE BALANCE IS]
  130
W: WITHDRAW 70
  [THE BALANCE IS]
  60
W: WITHDRAW 70
  [THE BALANCE IS]
  −10
```

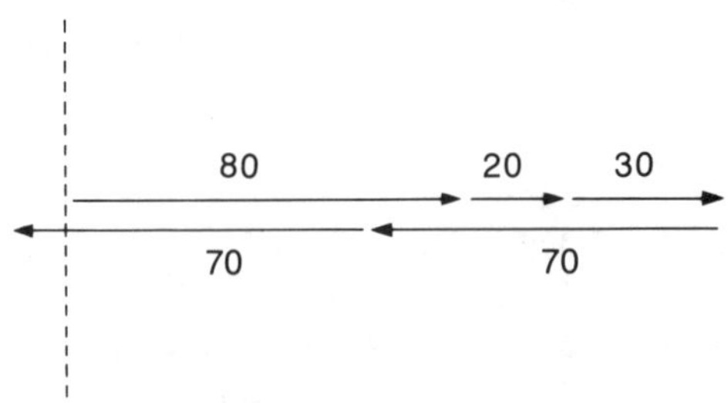

Fig. 6.35 — Money movements.

I also thought of this visual representation during the course of the session and, not realizing that Jane had adopted this solution, started discussing it with Mary. The recording shows how throughout a long discussion with

Mary, Jane was trying unsuccessfully to break into the conversation to ask how she could have her procedure print out the value of the plotter pen's position.

At the end of the session, Jane explained her insight:

> 'Well, I could not have done it with numbers. I find it much easier to use the plotter to do it and then go onto numbers.'

This showed how Jane used the explicit visual model of integers to solve a new problem.

In the next session subtraction of integers was presented as the inverse of addition. That is to say, the problem of solving the following equation:

$$(+230)-(-180)=X$$

was restated as

$$(-180)+X=(+230)$$

Jane then had to find the programming command which together with BACKWARD 180, gave a net effect of FORWARD 230. She found the correct answer, FORWARD 410, i.e. (+410). She was surprised that (+70)−(+160) could give the result (−90) because both the original numbers were positive.

When Jane had succesfully subtracted a number of pairs of integers, she was asked if she could formulate a rule for these subtractions. She was able to do this and explained a rule which was derived from the drawing work. This consisted of putting the directed numbers 'tail to tail' for subtraction, in contrast to putting them 'head to tail' as in addition, see Fig. 6.36. Although

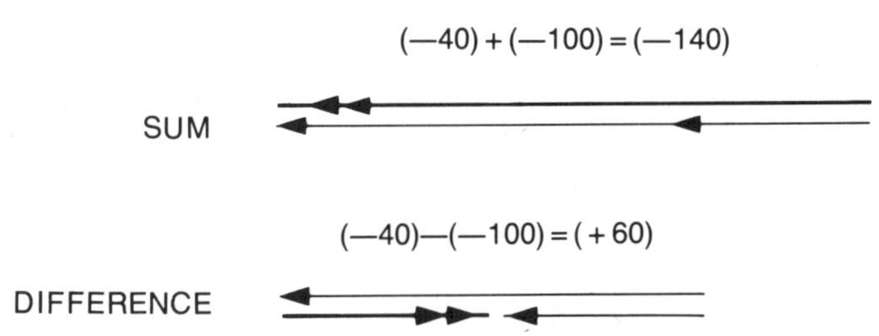

Fig. 6.36 — Combining integers.

it was Jane who thought of the rule initially, Mary explained it to me. She used the example (−40)−(−100).

Author: Explain to me how it [the new rule] works.

Jane: Well you start from the minus and you go backwards, well, it, you go backwards forty [draw (−40)].
Author: Ah, OK.
Jane: And then you go back again to the, your centre point there.
Author: OK.
Jane: And then you go back a hundred again [draw (−100)].
Author: Yes.
Jane: And...
Mary: And then you find...
Jane: Then you come back to...
Mary: From the second point to the first point.
Jane: That's right, u-hu, that's what I am trying to say.

What Jane had done was to construct diagrams similar to those often used to represent the subtraction of vectors, see Fig. 6.37. But she had done this

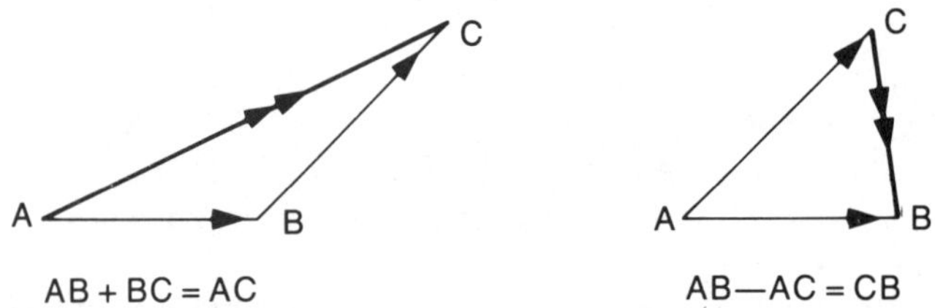

Fig. 6.37 — Combining vectors.

herself as a method of formalizing the drawing work which was modelling subtraction. One must add that the suggestion that such a formalism be sought was that of the worksheet, but the type of formalism was Jane's own.

Jane went on to explain about her earlier experience with integers:

> 'Yes, I can see now why, you know, you get pluses and minuses. But before when [the College lecturer] started talking about it, I didn't, you know... like he said 'a plus', like, 'a minus *and* a minus equals a plus'. You see, I thought a minus and a minus must equal a minus, and he was going spare because I said how does, how was it.'

Here she had misunderstood the lecturer's rule about multiplying integers by applying it to addition, possibly by misinterpreting the term 'and'.

Fibonacci series

While the previous sections have concerned topics with which Jane had earlier been observed having difficulty, this subsection describes her work with a number series. This work arose as part of the programming teaching on recursion. It is included here because it is anotherexample of how programming problems overwhelmed the mathematics.

The case in point concerns Jane's work with a recursive procedure which printed out a Fibonacci series. The definition of such a series was embodied in the recursive procedure (below) but the definition could be more easily stated as:

$$T(n)=T(n-1)+T(n-2)$$

where $T(n)$ is the nth term of the series.

The English statement of this is that 'each term of the series is the sum of the previous two terms'. In Logo, the procedure which generated a Fibonacci series gave an explicit description of how terms in the series could be computed, which obscured the simple statement of the generating rule. This would not necessarily be the case for other programming languages. The text of the Logo procedure was:

```
DEFINE "FIBONACCI "NUMA "NUMB
10 PRINT VALUE "NUMA
20 FIBONACCI (VALUE "NUMB) (ADD VALUE "NUMA
   VALUE "NUMB)
END
```

Jane spent some time puzzling out how the control structure and the binding of argument values in the recursive calls produced the given series of numbers. But she was not able to give me any concise statement about the 'rule' governing the series. Three months later, Jane returned to this problem but by then had forgotten how recursion worked so she could neither explain how the series was produced by the recursive procedure, nor could she give the rule governing the series. The next session Jane again spent some time on this problem and succeeded in explaining how the series was produced by the recursion saying:

> 'If I leave on a bad note [i.e. at the end of a session] it... you know... it preys on my mind all the time thinking about it.'

This work had involved her in considerable effort because of the complexities of the action of the recursion. But she still had little idea of the series rule. Three weeks later, I asked her if she could see the pattern in the sequence 0,1,1,2,3,5,8... which she had studied before. Now she was able to give the next term, 13, and to explain how she arrived at it.

There appear to be two difficulties. Firstly the series was introduced as an example of the action of a recursive procedure. It had not been requested by Jane and had no supporting mathematical context. Secondly the syntax of Logo obscured the rule rather than making it explicit.

SUMMARY

Jane's experience of learning mathematics through programming will be summarized using the framework set out in Chapter 5.

Rigour and explicitness

The evidence presented suggests that Jane had appreciated the value of the formal programming language, even if it was occasionally frustrating, because it enabled her to solve problems. Here the formality, though sometimes difficult to learn, was not seen as an expression of arbitrary rules but as a means of controlling a complex machine. Little work was undertaken to link the formality in programming explicitly to mathematical formality. But some pieces of conventional mathematical notation were explicitly linked to programming constructs e.g. parentheses and function notation.

Active exploration

Various modelling facilities were provided for Jane to explore different mathematical systems, e.g. Turtle Geometry, symmetry transformations, and fraction and integer operations. Numerous examples were given of Jane grappling with geometric problems in the course of her programming work. Though the notion of formal proof and theorems were not explored, a number of informal proofs and theorems were explored, for example, the total turtle trip theorem.

Instances were presented of Jane asking, if not answering, mathematical questions which arose naturally out of her programming activity. For instance, there was the pattern of the sum of the interior angles of polygons and there was the search for a relation between the angles of an 'arrow'. Sometimes the analysis, though grounded in the programming activity, moved beyond it. An example here was Jane's schematic diagrams for subtraction of integers. But much of her work concerned the solution of programming rather than mathematical problems.

Key concepts

Jane explored a wide range of important mathematical concepts including angles, functions, variables, integers. While the work with angle was extensive, not much was done with functions and variables. This was a shortcoming of the course of work rather than a difficulty associated with programming, or with Logo. The work actually done on functions showed that programming could be very useful. On a number of occasions it was necessary for me to make the explicit link between the programming work caried out and Jane's school-based mathematical knowledge. For example the relation betwen Turtle Geometry and use of protractors had to be discussed, as had the relation between the prefix function notation of her College worksheet and Logo's prefix procedure notation.

This suggests that, if students are to gain insight into key mathematical concepts, then their programming course should explicitly formulate links between programming and the student's existing knowledge of mathematics.

Problem solving

No attempt was made to observe changes in the students' ability to solve mathematics problems as a result of their programming work. Some evidence was gathered which supports the claim that programming provides a *language* for students to talk and think about their own problem-solving and other cognitive skills. For example, after Jane had solved the problem of drawing a house, she reported that:

> 'Oh, huh, its an awful lot of paper just to do that. I suppose you learn by... *by making the mistakes and having to put it all together again*... I think you do. It really makes you think about it.

In the context of programming the idea of building on one's mistakes took on a precise meaning. It was not just a pious injunction for approved problem-solving behaviour. Two facets of programming are important here, its explicitness and the printed record of the programming process. The explicit nature of the task made Jane really 'think about it'. That is, a sloppy procedure would either not run or would produce the wrong effect. If she was to get anywhere she had to think out exactly what she wanted the computer to do. The other point concerns the printed record of the session. Even though she did not examine it to make a detailed analysis of how she tried to solve the problem, she used it to make general remarks about the process, 'an awful lot of paper'. In principle, examination of these records by the subjects themselves could be used in much the same manner as videotape is used in micro-teaching sessions. The students did use these records of the session to see how a problem had been solved earlier, and occasionally to think about how a session had progressed.

In the programming classroom, Jane was a most conscientious worker who tended to stick to suggestions for work given in the worksheets. She was also persistent and would work at a problem for a long time. Normally I would leave her to get on with her own work unless she was obviously stuck, i.e. there was a long period of inactivity. Then I would go over to see what was wrong. I would also go over periodically to watch her and to talk to her about what she was doing. Jane tended to ask for help infrequently if I was not sitting with her. However, if I was sitting beside her, she would often let me take responsibility for what was done, asking how to do things which she knew how to do herself. It was difficult to steer a middle path between being too helpful and being too distant. There were also the further considerations of gathering research 'data' in the form of her comments on her work. This problem was accentuated by having only a few students working at a time, typically one or two. Thus there was often no-one else the student could turn to except me.

In the second questionnaire Jane commented on the way she had been taught programming:

> 'I think its a good idea to leave us to think things out *for ourselves*, at least for us to make an attempt, and then, if we're wrong, to go over, and find out where we went wrong.'

She was quite certain of the value of being left to solve problems herself. There was also an implied rebuke to me who, especially in the early sessions, used to 'hover' near the students and be too quick to give help.

In the third questionnaire Jane was asked what she considered 'the most useful thing in learning Logo'. She replied:

> 'Breaking down anything into smaller pieces especially in maths.'

This point was reiterated in her answers to other questions in the questionnaire. For example:

> 'I think that sometimes I can look at a problem and split it down into bits — before I tended to go headlong without really looking at a problem — great difficulties following!'

Thinking about learning

After the break during the summer vacation of 1976, Jane surprised herself because of the amount of programming knowledge she had retained. She explained why she was surprised:

> '... because first of all I have got a really hopeless memory for things. I find that... unless I keep going over a thing, I forget it very readily.'

She went on to explain that it was things like programming which she would normally have expected to forget. When asked why she thought she had remembered it, she ventured:

> '... possibly because you have got to think about it so much, and I suppose a lot of it does go in. Because you've got to... well like with this bit here [the work in hand] you have to, like when I was doing it before, you know, I had to keep going over it and over it again and you had to really puzzle out what when you made mistakes. I find that helps a lot when remembering it... and I make a lot of mistakes and I have really got to puzzle out where I've gone wrong and then I'll remember it better. It I had done it perfectly the first time, it would have been in and out [of my head] in two seconds.'

Jane explained at the end of the session that she had been expected to have forgotten everything and to have to go right back to the beginning. This fear was prompted partly by her belief in her own poor memory and also because of her experience after the previous large gap in her programming work, during teaching practice.

Two valuable lessons appear to have been learned. Jane clearly associated making mistakes, puzzling out those mistakes and learning and remembering. She also had a concrete instance of her ability to learn and retain this kind of subject matter.

She was also able to make some comparison between her experience of programming and that of a child learning mathematics. Thus from the second questionnaire we have:

> '... its something new for us as learning maths is for a child. Many problems will be similar. *But I came to Logo with a dread of maths*. That probably colours what I do — hopefully a child wouldn't feel like this.'

And later on in a session:

> 'Yes you do... I think you do... I think obviously you can appreciate a child's difficulty because when you come up against something there and it's really hard and you've got to work away at it... *and we find it hard, what must it be like for a child*?'

Here the value of the programming work was that it placed Jane back in the position of being a 'learner' again in a very obvious way so that she could get some insight into the kind of difficulty which might be faced by the children whom she taught.

Attitude to mathematics

A number of small increases in self-confidence were observed, related to Jane's increased understanding of particular mathematical topics, rather than a wholesale shift in her belief in her ability to cope mathematically. The evidence for this has already been presented in the sections devoted to each topic. Some of it will be briefly repeated here, followed by evidence of her attitude to mathematics in general.

(a) Geometry

> 'I can sort of visualize an angle... *but before it was dreadful*.'
> 'It's probably getting a bit easier now to think about angles. *whereas before it was just a haze*.'
> '... feel I could tackle this v. simply [in the classroom] — *before I couldn't have attempted it without a lot of help*.'

(b) Algebra

> 'I feel a *bit more confident about these ones* [composition of transformations]'

(c) Fractions

> '*I know what it means now* [equivalence of fractions], because I was never really clear exactly before.'
> 'If I hadn't done this [work with fraction procedures] and some child had asked me... I wouldn't 've... aha, I would say its just a rule.'

(d) Integers

> 'Yes, I can see now why you get pluses and minuses.'

(e) Attitude to mathematics in general

In answer to the second questionnaire, about whether she enjoyed using Logo and why, Jane answered:

> 'I enjoyed it at the beginning and still do. Why? Absorbing. (2) I can 'see' how some of the ideas can be applied to school maths (very simple thought) *which is great as before maths was just a fog*! (3) Pulls a lot of stuff out of your mind *that I didn't know was there*! *Obviously there's frustrating times, but, on the whole, I still enjoy it.*'

Her penultimate comment suggests that she has been pleasantly surprised by her ability to do programming. The 'frustrating times' she alludes to were when she was trying to solve a problem and also when the direction and value of the whole enterprise was in question. Thus during one session she described that sometimes the programming work had been like 'looking through a veil' and said that it had been a complete 'fog'. This seemed to be when the programming aspects of the work had overwhelmed the mathematical aspects.

In the third questionnaire Jane was asked if her attitude to mathematics had changed. She replied:

> '... slightly more confident, but don't think *anything* could make me feel completely confident.'

So it seems that the programming had helped a little, but she seemed resigned to the fact that mathematics was ever to be a cause of worry. She was also asked what she thought she had learned through Logo. She replied:

> '(1) Tackling problems more confidently. (2) Basic maths that I didn't know before. (3) To be more aware of difficulties in a topic, e.g. measuring angles.'

> It would appear that the work had helped her self-confidence a little. Her last statement indicated that she had become more aware of the difficulties of the children, but that she was beginning to be able to see causes for the difficulty which she could possibly tackle.

Disadvantages of programming

Although the programming work did provide Jane with many opportunities for doing mathematics, there were also many instances where programming hindered her mathematics work. These hindrances may be divided into two classes. One was caused by the ready availability of the machine for experiments and the consequent focus on production of programming 'products'. The other was caused by the complexities of programming itself especially where an inappropriate set of primitives was used.

There were many instances where Jane perceived her activity as merely the construction of a particular procedure or the production of a given drawing. This is inferred from the following frequent characteristics of her programming behaviour:

(a) Lack of overall analysis of the problem and little preliminary planning.
(b) Extensive use of successive refinement methods of procedure writing.
(c) Failure to search for a more elegant solution once at least one solution had been found.

One example is her work on tessellations. The procedures she wrote were not especially complicated, so programming complexity was not a problem. However, she concerned herself essentially with producing pictures of tessellations rather than with an analysis of tessellation. These pictures were built up by drawing a little of the picture, debugging it, and then drawing a bit more. There was little prior planning. Once the picture had been produced, there was little attempt either to search for a more elegant solution or to think about the underlying mathematical constraints on the tessellation properties of different regular polygons.

Another example was the unsuccessful fraction project. Here Jane had a plan of attack, but her plan obscured the mathematical point which the procedure was intended to illustrate. This happened because Jane attended to the more obvious problem of how to draw the fraction pie-chart, rather than to the underlying problem of what such a pie-chart represented.

These difficulties must be attributed to the way Jane was taught to program and to the kind of question the mathematics worksheets asked. They illustrate a very real difficulty which faces those who wish to design mathematics courses based on programming, that is, the design of programming projects which successfully confront the student with mathematical issues without overwhelming her with irrelevant programming detail.

A second way in which the programming work hindered mathematical insight was caused by the complexities of the programming language. Here an example was the work done on Fibonacci number series. Most of Jane's effort went into finding out how the recursion worked and how the given series was produced by the procedure, rather than into understanding the series generating rule. This difficulty might have been reduced if Jane had studied a number of different series-producing procedures. Then once recursion had been understood, she might have thought about the properties of the series themselves.

Many of what appear to be the most productive sessions involved Jane either writing very simple procedures or running procedures provided for her. These new primitives enabled her to study mathematical topics at an appropriate level of representation.

The programming sessions gave Jane the valuable opportunity to reveal and explore some of her mathematical difficulties. She succeeded in understanding some of the topics which had been puzzling her (for which she had been unable to get help within the College of Education). The practical benefit of the computer was that it enabled her to explore mathematical topics herself and experience the joy of success.

7

Irene and Mary

The two case-studies presented in this chapter describe incidents from the work of Irene and Mary. These case-studies are less detailed than Jane's and only include incidents which contrast with Jane's experience.

Irene found programming difficult and unpleasant. She never mastered it sufficiently to get much benefit from it. She took a passive attitude towards her programming work and initiated few programming projects, but was content to work at what was suggested. She concentrated on the programming issues of her work and hardly ever explored the underlying mathematics.

Mary was quite different. She was much more curious about mathematics than either Jane or Irene and she used programming to explore mathematics to a much greater extent than the other two students. Not only did Mary notice mathematical patterns, as Jane had done, but she constructed explanations for those patterns. Her case study demonstrates that, once a student has understood a number of computational terms, these may be used as a vehicle for mathematical explanation without the necessity of running programs. It was sufficient to explain concepts in terms of hypothetical programs.

7.1 IRENE

Irene's case study concentrates on three aspects of her work. First of all she found programming and Turtle Geometry confusing and frustrating. Secondly she adopted inefficient problem-solving methods (such as trial and error, and 'linear refinement') and concentrated on producing pictures rather than on understanding the geometric properties of the pictures. Thirdly she was given a poorly designed programming project which attempted to help her with a classroom difficulty concerning division.

Irene spent about 27 hours in the programming classroom. A table of her month by month programming activity is given as Fig. 7.1. The immediate

Fig. 7.1 — Irene's programming sessions.

contrast with Jane was in her relationship to the programming work. While Jane was conscientious, missed few programming sessions and telephoned if she did miss any, Irene had what appeared to be a much more casual attitude. She sometimes missed sessions but would come with excuses to a later session. She revealed later that one of the reasons why she missed sessions was because she found the programming so difficult. Missing sessions had the effect of making programming even more difficult for her to grasp. As a result Irene learnt less programming and covered far fewer mathematical topics than Jane. Unfortunately Irene came to many of the programming sessions on her own, whereas Jane and Mary often came together. This increased her difficulties because she did not benefit from the help of another student.

Frustration

Most of Irene's programming was carried out using the drawing devices. Irene, like Jane, had difficulty defining a procedure for a triangle though she did not make exactly the same kind of interior/exterior angle mistake as Jane. Irene tried to draw an equilateral triangle with three exterior angles of 45 degrees. This had the effect of drawing part of a regular octogon. She corrected this after a discussion with me about 'walking around a triangle', the total angle turned and the angle turned at each vertex. Her difficulty over the way the turtle turned reappeared throughout her programming work.

After teaching practice, Irene had difficulty adjusting to the changes in the Logo implementation. This meant that much of her effort was concentrated on formulating syntactically correct commands rather than on exploring mathematics. At this time she filled in the second questionnaire which asked about her programming experience. Irene wrote:

> 'Yes I did and still do enjoy using Logo. I find it fascinating yet confusing therefore I feel I have still got a long way to go before I get full meaning/understanding and benefit from it.'

Her mixed feelings about programming, 'fascinating yet confusing', were expressed with more force on further occasions. Her answer to the question about how the 'rotated house' picture could be drawn was much less explicit than Jane's, though Jane had the advantage of having just solved part of the problem prior to answering the questionnaire. Irene's answer was:

> 'Find a starting point, (turtle)→ connect and they [sic] begin giving Logo various procedures.'

Irene had a lot of difficulty when she came to draw a 'house' soon after. The session started with Irene listing various procedures she had defined in her first term including a procedure for drawing triangles of any size. When she started on the problem of drawing the 'house' these procedures were ignored. Again she defined a triangle procedure with angles of 45 degrees instead of 120 degrees as she had done before. By running the procedure she saw that it was incorrect. I have her some help in the method of editing procedures and then asked her about the angles of the triangle. She said that they had to be less than 90 degrees because it was an equilateral triangle. Irene asked if the triangle in the worksheet was meant to be equilateral. I said that the illustration in the worksheet might have been a little inaccurate because I had drawn it with a ruler. Irene retorted:

> 'I wish you would just get me a ruler, so that I could just draw with it too.'

Her comment indicated a little of her frustration. She seemed to regard the programming as a means of producing a given picture 'product'. In this respect programming seemed highly constraining because it severely limited both what could be done and the means by which it could be achieved. From Irene's point of view the precise and rigorous formulation of procedures to draw the house seemed rather pointless, especially as I had produced a similar picture with ruler and pencil for the worksheet.

I continued my questioning about the angles of the triangle, asking whether they should be more or less than 45 degrees. This time Irene said she did not know, but added that she remembered being told about the angles of an equilateral triangle before. She also added, a little accusingly:

> 'You're not telling me anything, are you?'

I explained that I believed it better that she experiment for herself because I had told her the answer before, and she had forgotten it. She then asked:

> 'It's not something about three hundred and sixty or was it hundred and eighty?'

I advised her to try a value different from 45 degrees even if it was just a guess. Eventually she suggested 120 degrees but could give no proper explanation. She said that an 'obtuse' angle was greater than 90 degrees but that she had been trying to draw an 'acute' angle. This was why she had tried values less than 90 degrees.

> 'It seems that, for to get it on that [the display turtle] . . . to get an acute angle, you've got to do more [give a number larger than ninety].'

I then suggested she work with the button box and floor turtle because experiments could be conducted faster and because the rotations of the floor turtle were clearer than those of the display turtle. I asked her to draw a hexagon. She tried turns of 15, 5, 75 and finally 60 degrees in her attempts to produce a regular hexagon. She was beginning to get cross and so I gave her more help by describing the total turtle trip theorem and the importance of 360 degrees. Irene spent the rest of the session trying to fit her triangle (now debugged) and square together to form a house using the display. She found it very frustrating, saying 'I am getting exasperated'.

I suggested various tactics like clearing the screen and re-positioning the turtle at the centre of the screen between each of her attempts. Eventually she abandoned her search for the correct set-up step between the square and the triangle, and decided to go home. I discussed the session with her:

Author: Has it upset you?
Irene: No, I'm just annoyed.

I tried to reassure her that she had learnt quite a lot and should come again soon to finish off the outstanding problem.

Irene: . . . I just get myself into a. . .*I feel so stupid. . .I know. . .I really. . .it makes me feel like, och, I just feel really thick.*
Author: Well that's, I mean, that's a shame. I don't want you to feel that way about it.
Irene: *Em, my mind goes exactly the same way as it does in maths, exactly the same.*

Unfortunately the programming work had put her into exactly the same frustrating and debasing position she remembered from mathematics. She expanded on this point:

Irene: I put on the end of that thing [the second questionnaire], I hope by the end of this course [the programming work] that my attitude towards mathematics will be one of pleasure and [she laughs] enjoyment.

Author: But going by today's thing you are not terribly hopeful about it.
Irene: Och yes, I enjoy it.

Irene's avowal that she 'enjoys it' seemed to contradict her earlier comments, 'I feel really thick', 'so stupid' and 'my mind goes exactly the same way as it does in maths'. This contradiction will be explored presently.

Drawing pictures

Irene returned within a few days to finish solving the problem of drawing the 'house'. This time Jane was also present and was able to give her some help both with the language and with the drawing problem. Irene used the floor turtle and eventually, after trying various commands to orientate the turtle between the square and the triangle, she found the correct angle. Her search had been prolonged because the inaccuracy of the floor turtle masked the true solution. This led Irene to make two unnecessary attempts which differed from the true solution by 5 degrees. Her strategy had been predominantly one of visual inspection of the turtle's movement rather than an analysis of angles. When she had completed the drawing she did not know why the angle she had chosen was *necessarily* correct. I helped her establish this by working through the procedure and by matching the turtle's movements to the individual commands.

Although she had drawn the triangle with the appropriate angle of 120 degrees, she still thought that its interior angles were 45 degrees. She seemed convinced that an equilateral triangle had angles of 45 degrees. I explained that the angles inside an equilateral triangle had to be 60 degrees, given that she had used 120 degrees as the correct exterior angle. However, the point was not pursued because she was pleased with her picture and was not much concerned to establish why it worked.

Irene's lack of analysis and her use of trial and error methods of solution diminished the mathematical benefit of the task. Irene saw her task as the production of given drawing while I wanted her to examine the angle properties of the figures drawn.

Next session, Irene explained that she was pleased to have solved the 'house' problem and that she had put the picture of the 'house' up on her wall at home. She had shown her father the picture and the teletype record of the sessions.

> 'My dad sort of looked at it [the picture] and then looked at the long piece of paper. 'My God,' he said, 'you doing all that just for that wee thing.' You know they think it's all simple. But it's not really simple, not to work out how to get it. I don't think it's simple.'

Like Jane, Irene had to reconcile the apparent simplicity of the 'product' with the complexity of the process of achieving it. The remarks of Irene's father and those of Jane are remarkably similar. Jane, however, saw some value in this process where Irene seems only to have seen the complexity.

About four months later, after the summer vacation Irene attempted further simple drawings. She wished to draw a diamond which she planned as a pair of triangles, see Fig. 7.2. Her initial attempts were hampered by a

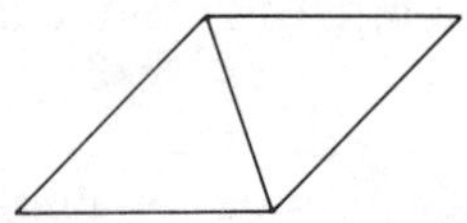

Fig. 7.2 — Diamond.

system crash and so she was only able to produce a single non-state-transparent triangle. She made mistakes because of the long gap since she had last programmed and confused procedure execution with procedure definition. She returned to the problem at the next session. This time her initial attempt produced part of an octagon. Then followed a long debugging session at which she adopted an inefficient 'linear refinement' strategy. Once the 'diamond' was debugged, Irene started to incorporate it, as a 'leaf' and a 'petal', in the drawing of a 'flower' as suggested in a worksheet. The session was discussed:

Irene: *It's a funny thing I think that you, you [herself]get a wee bit, a bit excited.*
Author: Yes.
Irene: You know then it turns out somehow wrong and. . .

Irene seemed surprised that she could get 'excited' about such an activity. She explained that:

'My brain isn't logical, that's why I don't get on with this.'

She continued:

Irene: . . .still just *it makes you feel so stupid,* you know, it's incredible.
Author: Well I didn't, I didn't intend, I don't intend that.
Irene: Well, I mean, I suppose. . *.it's good anyway. It does make you. . .it really does make you really think.* But I don't get. . .'

She had expressed these sentiments before about Logo making her feel 'stupid'. The programming work was confirming her belief in her own inability rather than eradicating it. Her comment that 'it's good anyway' was uttered without much conviction, in rather the same spirit that one might express oneself about some strenuous and unpleasant exercise prescribed by a doctor.

Irene had mixed reactions to this session. She was both repelled and attracted by programming. She believed that programming was good for her but found it very hard and so felt 'stupid'. This feeling was, no doubt, accentuated by being observed by me whom she knew could solve the programming problems. Her earlier surprise at getting 'a wee bit, a bit excited' makes sense as her comment on her *unexpected* reaction to what was usually a rather unpleasant activity.

Irene continued to debug her 'flower' in the next session. She was adamant that I should not stay near her, watching her. I adopted the policy of working elsewhere and returning at intervals to see how she was progressing.

Irene found the debugging difficult because she had no clear plan and was not sure which part of the 'flower' each of the procedures was intended to draw. Her habit of defining a new procedure, with a new name, as a method of debugging an existing procedure had produced a plethora of procedures with similar names and functions that was thoroughly confusing. I helped her by getting her to explain what each procedure was to do and what the net effect of each procedure would be on the state of the turtle. Eventually with more help a 'flower' was produced and Irene asked: 'Will that do?' Irene seemed quite glad to have finished it. She seemed to look on this programming problem as an *imposed* task which it was her duty to complete to my satisfaction. Irene was more concerned with the picture 'product' of the session than with either understanding why the program worked, or with constructing personally pleasing patterns. By way of comment on her performance, Irene noted that:

> 'My brain needs a good shaking just now'.

Failed programming project

Irene's classroom difficulty with division had been observed earlier. This concerned the distinction between sharing out objects among a given number of people and partitioning a set of objects into sub-sets of a given size. I explained the distinction to her while she listened to the recording of the lesson in which the difficulty had arisen.

Irene was given a programming project which described the action of procedures to exemplify the distinction between sharing and partitioning. The project was unsuccessful because Irene found the procedures too difficult to write. Also the illustrations provided by the procedures (which were explained by me) were not clear enough to characterize the distinction. Both procedures took argument values. The procedure illustrating 'sharing' worked as follows. Here the number of objects (12 and the number of recipients (4) was known:

```
W: SHARE [T T T T T T T T T T T T] 4
[T T T]
[T T T]
[T T T]
[T T T]
```

This indicated that each of 4 people received 3 objects. The procedure to illustrate 'partitioning' worked as follows. Here the number of objects (12) and the size of the sub-sets [T T T] was known.

```
W: BREAKUP [T T T T T T T T T T T T] [T T T]
4
```

Irene would probably have benefited more by manipulating real objects rather than by attempting to write procedures which manipulated symbols representing objects. The former would have been closer to her classroom experience with the children and would have revealed the distinction more clearly.

I was able to re-observe Irene teaching division. My explanations and the partially completed programming work produced only a small change in Irene's approach to teaching division. The children seemed to muddle sharing and partitioning more than ever.

Irene issued straws to the children. When the straws were shared out by the children, some counted the number of straws each had got and others counted the number of groups of straws produced by the sharing.

Irene: Now I said, what did I say you were to do, David?
David: Share out so each person has three.
Irene: Each person has three, no. We are going to share out sixteen straws to three people and I want these three people to have exactly the same amount of straws. So what, well what. . .
pupil: They will all get four.
Irene: They will all get four, no. You've got five, you've got five and you've got five and there's one left over.
pupil: *Three remainder one.*

The pupil counted Irene's three utterances of the phrase You've got five' and looked at the three groups of straws on the table. This pupil's answer precisely characterized the partition/quotition muddle.

Irene: Three remainder one, no. Why do you think it's three remainder one?
pupil; [inaudible]
Irene: *Three bundles. You've got how many. . .in those bundles.*
pupil; It's five remainder one.
Irene: Five, it's five remainder one.

At the end of the dialogue, Irene did direct the pupil's attention to the number of straws in each bundle as opposed to the number of bundles.

However, she did not make the point with any force. Irene's instructions still contained ambiguities, though she did not mix sharing terminology and partitioning terminology as strongly as she had done in the first lesson observed.

7.2 MARY

Mary gained much more mathematical benefit from her programming work than the other two students. She was mathematically more competent than they were. She scored higher in the mathematics test and displayed fewer difficulties in the classroom. Her case study shows how she used the drawing devices to create her own pictures. Like the other students, she wondered how learning to program would help her as a teacher. But she conducted a successful investigation of the mathematics of Turtle Geometry and made a number of personal mathematical discoveries. She used the computer to test hypotheses and as a source of interesting mathematical problems.

Mary studied functions by considering the code of a set of hypothetical procedures. She did not need to run these procedures, but she just hand-traced through them. This helped her to disambiguate the conventional function notation employed by the College of Education. She brought mathematical questions which stimulated her to the sessions. Her interest in them was fuelled by mathematical curiosity rather than by specific teaching needs. In many ways Mary had the least need of mathematical help and benefited the most from it.

Mary spent about 50 hours programming. Like Jane she missed few sessions and covered a wide range of mathematical topics. A chart of her month-by-month programming time is given as Fig. 7.3.

Fig. 7.3 — Mary's programming sessions.

From the start Mary worked quickly and self-confidently. Initially she was worried about asking questions but soon realized that I welcomed them. Unlike Jane and Irene, she often noted down my replies and would often explain my answers back to me in her own words. Where Jane and Irene would listen passively, Mary would interrupt, ask further questions or summarize what had been said. Her whole approach to programming was more active. She was much more likely to mention bits of mathematics which either confused her or *interested* her. Of the three students, she was the most curious about mathematics, both for its own sake and because she needed to understand it in order to teach effectively.

Drawing pictures

Mary had little apparent difficulty with the interior/exterior angle distinction and could produce correct turtle drawings more easily than either Jane or Irene. After a long break, caused by teaching practice and by her College examination, she coped with the implementation changes, asked lots of questions and was easily able to define a procedure to draw an arrow.

Mary answered the second questionnaire when she started programming again in June. Like Irene, her answer to the question about how the 'rotated house' should be drawn was not as explicit as Jane's:

> 'You will need to get the turtle or other drawing devices. Give the commands. Ask Logo to define the commands.'

When she came to solve the problem of drawing a single 'house', she had difficulty with the distinction between procedure execution and procedure definition. She correctly used the angle 120 degrees as the angle to draw the equilateral triangle. She spent some time trying to fit her square and triangle together. Her difficulties were made worse because her procedures were not state-transparent and because she lifted the display turtle's pen between drawing the square and the triangle and forgot to put it down again before drawing the triangle. Eventually she produced a 'house' whose commands she then embodied in a procedure.

Mary then went on to use the 'house' procedure as a sub-procedure to draw a 'street' of 'houses'. In the debugging sequence for the 'street' she inadvertently produced a pattern of two 'houses' at right angles, as in Fig. 7.4(a). She saw the possibilities of this and deliberately continued the pattern, see Fig. 7.4(b)).

When the pattern was complete she defined a procedure, named SIGN, which drew a quarter of the pattern: a 'house' followed by a translation and a turn ready to draw the next 'house'. She then drew the whole pattern again on the screen by running SIGN four times. Then Mary returned to the problem of drawing a 'street of houses'. She realized that her procedure SIGN solved the problem of putting 'houses' together if she undid the effects of the turn command which rotated them. She successfully drew a row of four 'houses' using the sequence in Fig. 7.5.

Mary then defined a procedure named STREET consisting of the pair of

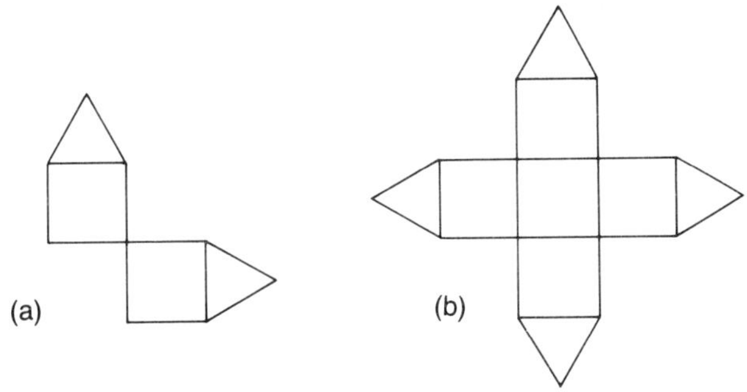

Fig. 7.4 — Pattern of houses.

```
SIGN
LEFT 90
SIGN
LEFT 90
SIGN
LEFT 90
SIGN
```

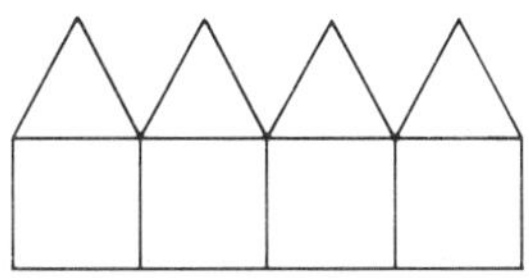

Fig. 7.5 — Street.

commands 'SIGN, LEFT 90' and spent the rest of the session drawing 'streets' of different numbers of 'houses' all over the screen.

Mary's behaviour in this session was in strong contrast to that of Irene who had also tried to draw a 'house'. Mary said that she liked to ask questions and did not mind whether I was watching her. She worked faster than Irene and made many mistakes. But she seemed much better at recovering from her mistakes herself, whether they were syntactic or geometric. She was much more personally involved with the work having produced an interesting pattern as a by-product. She also went much further than Irene who had to struggle to produce even a single 'house'.

Next session, Mary attempted to define a 'flower' following a suggestion in a worksheet. One of the procedures which she had to define was that for a rhombus (a 'diamond'). Her first attempt, like Jane's, had three angles of 45 degrees. However, she immediately diagnosed the error and fixed it. But she did not succeed in completing the 'flower' because the super-procedures given in the worksheet expected her 'diamond' sub-procedure to be state-transparent which it was not. After spending some time attempting to draw the 'flower', she abandoned the problem and instead defined a long fixed instruction procedure which drew a 'range of mountains' similar to Fig. 7.6.

Mary's reaction to programming and its setbacks was in marked contrast to Irene's frustration and feelings of incompetence:

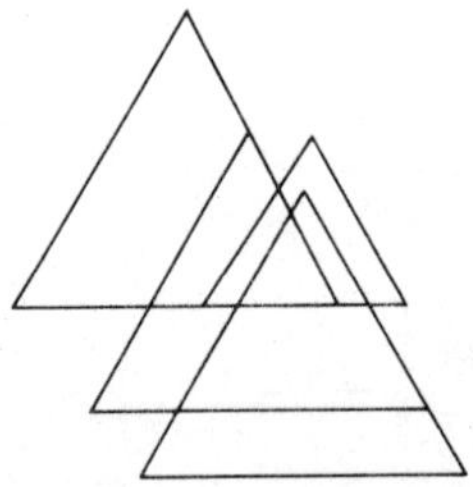

Fig. 7.6 — Range of mountains.

> 'When you go to Logo you never realize how the time flies. . .and you forget about all the other worries'.

Author: Did you enjoy the session today?
Mary: Yes I think I did. It was, eh, no especially this achievement, that was good, you know, where *I could find my own mistake and where I could correct it at the same time.*

Mary had become quite adept at drawing with the turtle though she still completely misunderstood procedures which took arguments. In her session she defined a hexagon procedure and went on to use this as a building block in a complex picture of a 'railway bridge', see Fig. 7.7. Again she was pleased with the session because she had solved a variety of problems herself, though the difficulty over arguments had not been resolved.

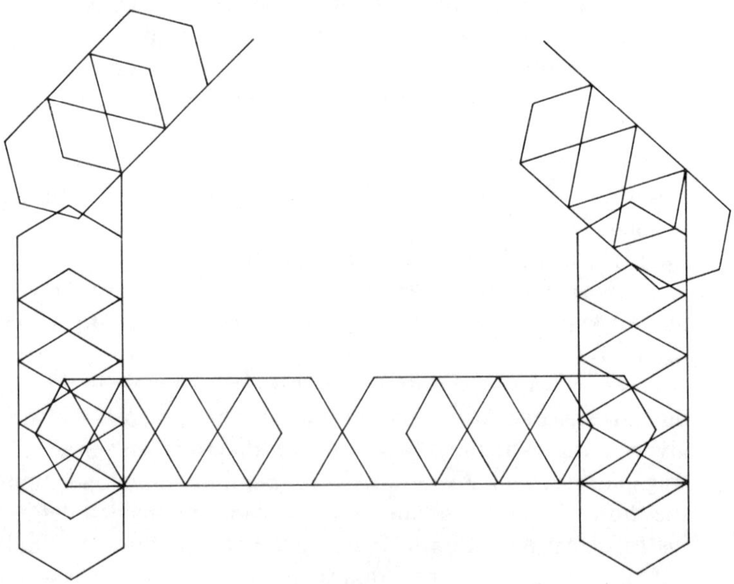

Fig. 7.7 — Railway bridge.

Doing Turtle Geometry

The following sessions are described in some detail. They show how Mary's interest in the properties of polygons was stimulated by her programming work. They also show how she explored and discovered relations among the angles of regular polygons. At the start Mary knew the sizes of the angles of familiar regular polygons such as squares. By the end she demonstrated an understanding of how squares shared a number of properties in common with other regular polygons. These sessions indicate that it is possible for students to explore Turtle Geometry to gain mathematical insights.

Mary drew a square and a pentagon whose angles were given in the worksheet, and she was able to draw an octagon, having calculated its external angle herself. She tried to draw a 36-gon (exterior = 10 degrees) but it went off the display screen and so she could not see whether it closed or not. She correctly drew a 12-sided polygon (a dodecagon: exterior angle = 30 degrees) and then tried a second 12-sided polygon with an exterior angle of 20 degrees which was too large for the screen. This polygon would not have closed. She wondered why this last polygon was too large for the screen and then explained that a polygon with an angle of 20 degrees would be 'broader' than one with an angle of 30 degrees and so would not fit on the screen.

Author: Do you know how to decide whether the thing [polygon] is going to close up or not?
Mary: Er, no, this is what I was trying to discover, in fact.

I suggested that she explore a bit further to establish the closure rule. But she asked how many degrees the turtle turned in a closed path. I then showed Mary that a body rotated through 360 degrees when it moved around a closed path which did not cross itself. I asked her what angle it would turn through at each vertex of a square.

Mary: It is ninety. . .ninety four times.
Author: Right. . .ninety four times, yes OK, which comes to?
Mary: Three hundred and sixty.
Author: OK right. A pentagon?
Mary: That's five, er, three hundred and sixty divide by five.
Author: Right, OK, that's it. That's the rule.

Mary then examined her 12-sided polygon with 30-degree angles, and her attempted 12-sided polygon (partially off the screen) with 20-degree angles and said.

> 'Aha, so twelve twos are twenty-four. . .so two hundred and forty and so that wouldn't close up.

Mary's response to the worksheet instruction 'draw a lot of different polygons' was to try to establish a rule for closure. Where Irene had taken a similar instruction quite literally and drawn some polygons, Mary concerned herself with the properties of the polygons. It seemed that now Mary understood the relation between the number of sides and the exterior angles of regular polygons. So I suggested that, for homework, she think about how a circle, a five-pointed star and a six-pointed star might be drawn, see Fig. 7.8. Mary found this a stimulating problem and spent much time wondering

Fig. 7.8 — Two stars.

about it before the next session.

'I think I worked over it for three hours. . . .'

When she arrived at the session, Mary immediately mentioned the 'triangle within a triangle', the six-pointed star. She drew one by hand and labelled certain angles, incorrectly, as shown in Fig. 7.9. She reasoned that since

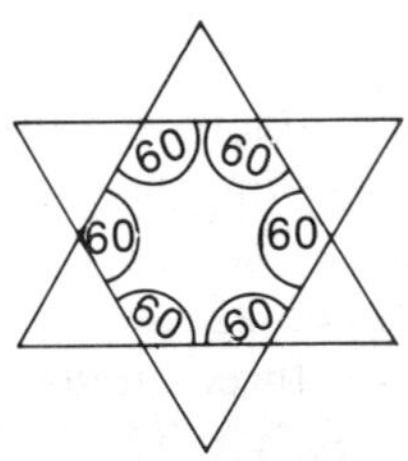

Fig. 7.9 — Incorrect angles.

there were six angles, each had to be 360/6 = 60 degrees. This was a misapplication of the total turtle trip theorem which she had explored in the previous session. Mary explained that her labelling led to a contradiction which she was about to describe when I interrupted her to point out that she had labelled the angles incorrectly. I explained that the 360 degree total referred to the total exterior angle rather than to the total interior angle.

Mary did not see the force of this until I asked her whether the angles she had labelled as 60 degrees looked more or less than 90 degrees.

> 'Bigger than ninety degrees. . .aha. . .ah, now this is where. . .this is what I couldn't solve it, you see. . .and I didn't have a protractor. . .so I couldn't measure it. But I went and asked somebody at the office, I mean, where I am working at the moment [a vacation job] and he was thrown as well because. . .so if that would be, that is ninety [she estimates the size of the angles], that would ninety. . .that would be forty-five degrees, eh, ah so that will be hundred and twenty degrees.'

I again pointed out that the total *exterior* angles of any polygon which does not cross itself is 360 degrees but that the total *interior* angles depended on the particular polygon. Mary told me the total interior angles for a square and a triangle correctly. I started to draw up a table, see Fig. 7.10, with Mary's help. She now saw her mistake and was able to anticipate many of the entries.

No. of sides	Interior angle	Total of interior angles	Exterior angle	Total of exterior angles
3	60	3*60 =180	120	3*120=360
4	90	4*90 =360	90	4* 90=360
5	108	5*108=540	72	5* 72=360
6	120	6*120=720	60	6* 60=360

Fig. 7.10 — Table of polygon angles.

> 'Because I asked, and they [the office staff] didn't know, you know, and so they said 'well you are a teacher, you ought to know'. . .and I was trying to puzzle that out. . .and I couldn't because I didn't have a protractor but it was stupid of me, I mean, I didn't look at the size of the angle.'

Mary then filled in the table for the square and the regular pentagon. For the pentagon she calculated the exterior angle first and then the other angles, following my suggestion.

Mary then noticed that the interior angle total increased by 180 degrees for each extra side. Jane had noticed the same pattern but had been unable to explain it or make use of it. I suggested that Mary see if 'her law' continued to work for polygons with larger numbers of sides. She established that it did and happily announced that:

> 'I have discovered something new, yes'

Mary was then able to give an explicit method for calculating all the angles of any regular polygon:

Mary: . . .a regular one, you would say that eh, what eh, number of [sides]. . .that is *x*, multiply by. . .no, sorry. . .*x*. . .three hundred sixty divide by *x*, ah, that would give you the exterior angle [exterior angle = 360/number of sides].

Author: Yes.

Mary: And eh, that multiplied by the number *x*, I mean no sorry.- . .hundred and eighty take away the number you got would give you the interior angle [interior angle = $180 - 360/x$] and if you wanted to find the total number of interior angle. . .that means one interior angle multiplied by the number of sides, that is *x* [total interior angle = $x(180 - 360/x)$].

Mary then returned to the contradiction which I had interrupted earlier. She reasoned that if she labelled the angles as in Fig. 7.9 it would imply that the angles in the triangle in Fig. 7.11 would total at least 240 degrees but:

> 'Funny, two hundred and forty but a triangle has only *got* hundred and eighty degrees.'

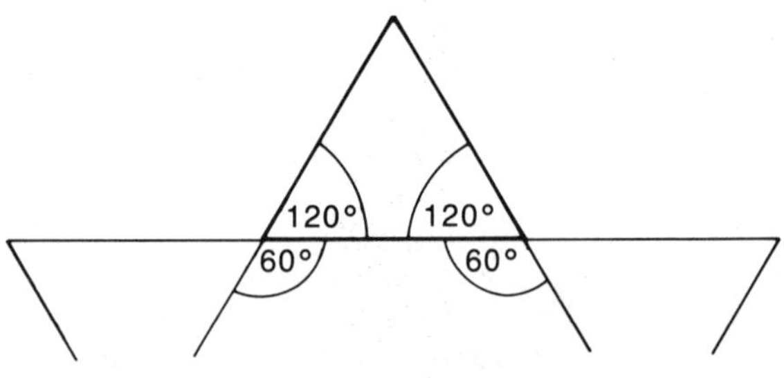

Fig. 7.11 — Incorrect angle values.

This was the contradiction that she had been unable to resolve. Mary then mentioned a difficulty in connection with circles. The basis of the difficulty was that she had not thought out clearly what she meant when she asserted that a circle had 'got' 360 degrees. The difficulty was similar to her misunderstanding about the hexagon in the six-pointed star which she then thought had 'got' 360 degrees. She now knew that a hexagon had a total of 720 degrees for its interior angles. She said that if one then inscribed a hexagon inside a circle (see Fig. 7.12) it 'changed' the number of degrees within the circle from 360 to 720. I suggested that she might explore this

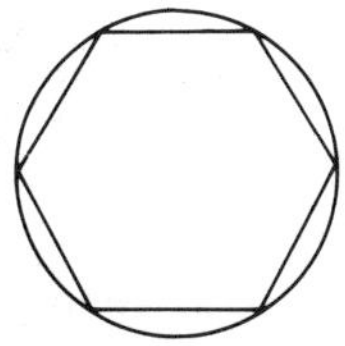

Fig. 7.12 — Hexagon within a circle.

misunderstanding by trying to draw a circle using her polygon procedure. However, she was very keen to draw the five-pointed star and decided to tackle this first and leave the circle until later.

Mary did not apply her knowledge of polygons to the star problem. First she defined a procedure for the star which had unequal lengths of lines and unequal angles, which drew a shape as in Fig. 7.13. She attempted to close the star by gradual modifications to the lengths and the angles in her procedure. She succeeded in completing an irregular closed star but was not satisfied with this as a solution. Her behaviour here contrasted with that of Irene and Jane, who were usually satisfied once they had produced the picture they were drawing. I asked her if she was trying to draw a regular star

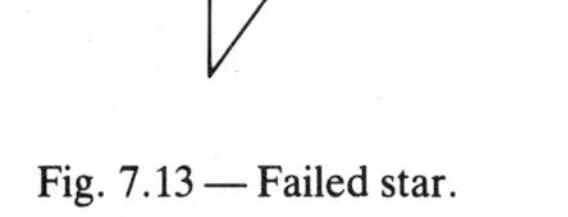

Fig. 7.13 — Failed star.

with all its lines the same length. She said that she was attempting this, although the lines in her procedure were of different lengths. A little later she had an insight into the problem:

> 'Oh, you know what, these angles in the middle should be the same throughout.

But she thought that the exterior angle total should be 360 degrees as it has been for the convex polygons considered earlier. I showed her how 'walking around the star' took two complete revolutions, not one. She then was able to calculate the appropriate exterior angle (720/5 = 144) and drew the star using her general polygon procedure. She was pleased because now it was much clearer to her than the programming work was 'getting into real mathematical things' whereas before she had wondered what the point of it was beyond teaching 'logical' thinking.

Finally Mary considered the 'interior' angles of a circle. I showed her that if she considered a circle as a polygon and followed the rules that she had already established then a circle had an infinite total for its interior angles. Mary tested this idea by drawing a 30-sided polygon which gave a reasonable approximation to a circle on the display screen.

Mary attended the next session with Irene. During a lull caused by a system crash, Mary told Irene about the polygon work from the previous session. Mary was very pleased to have solved the problem.

> '. . .there was a problem which I was supposed to solve and all of a sudden in the middle of the night it struck me why wasn't it working, you know. So I asked my sister, she just said 'Come on, throw that away in the middle of the night'. So I went to work in the morning and I asked the boys about it and one of them started playing a joke and said 'Why don't you ring to the Israel Embassy [about the six-pointed star] and they will tell you how to solve it. . .so I didn't get that solved, and then I came back here and I asked Ben [the author] about it. So eventually we got into this problem and we were able to solve it.'

Then Mary showed Irene that the total of the interior angles of a polygon depended on the number of sides. Mary then discussed the polygon angle table with me. She said that the total interior angle increased by 180 degrees for each extra side. She examined the sequence of individual exterior angles: 60, 90, 108, 120 etc. She asked:

> 'But I was wondering how much this [individual exterior angles] was going up by.'

I did not give her an explanation except to say it was an interesting problem. Mary continued to examine the table. Later she explained how she could use her rule, that the total interior angles increased in steps of 180 degrees, to calculate the angles of any regular polygon given the number of its sides.

Three weeks later, Mary suddenly mentioned the polygon table again. She re-explained that one could find the total interior angle of a decagon, say, by two methods. One method was to compute how many more sides a decagon had, compared to a triangle and then to add 180 degrees for each extra side to the interior total angle of the triangle.

Total interior angle = [(10 − 3) + 1] * 180

The other method was to divide the total exterior angle, 360 degrees, by the number of sides, 10, to find a single exterior angle. The interior angle was the supplement of this, and the total interior angle was a single interior angle multiplied by the number of sides.

Total interior angle = (180 − 360/10) * 10

I asked Mary why her first method worked. At first she justified it in terms of the pattern she had discovered in the polygon angle table, but I persisted:

Author: Yes, I know that, but why, but why is that when you add one more side you get exactly a hundred and eighty more degrees in the interior?

Mary: [after a pause] Oh is it because it makes, it makes an extra triangle?

I suggested that she draw some polygons to make the point more explicit, which she did using paper and pencil, see Fig. 7.14.

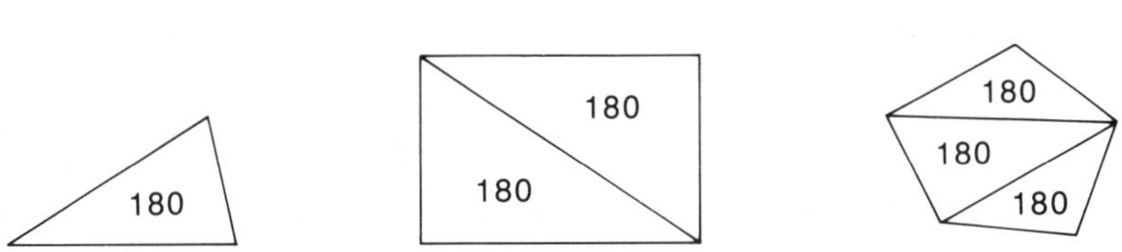

Fig. 7.14 — Adding triangles.

Mary: . . .so it is going up by each triangle.

The next session, Mary discovered a further property of regular polygons. She had defined procedures which repeatedly translated or rotated a basic shape such as a pentagon. She then considered the symmetry of various shapes such as the capital letters of the alphabet and the regular polygons. She explained that she understood reflective symmetry (because she could imagine a mirror placed over the figure) but found rotational symmetry harder to understand.

I suggested that Mary consider the problem of an equilateral triangle in an identically shaped frame, see Fig. 7.15. Could the triangle be taken out, rotated in its own plane and then replaced back in the frame? Mary suggested that a 360-degree rotation would work and then saw that a smaller rotation would also do:

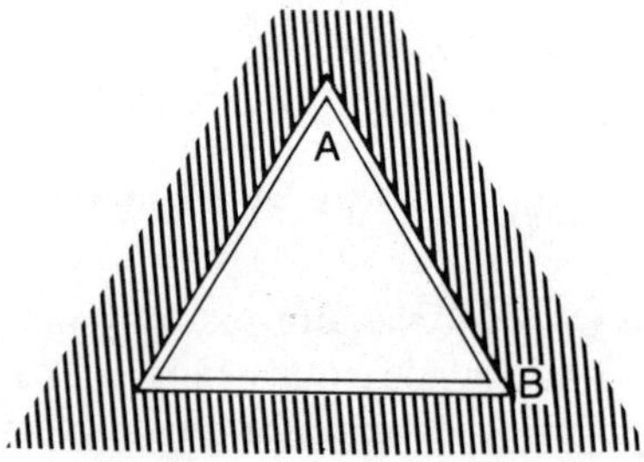

Fig. 7.15 — Triangle in a frame.

Mary: . . .if you took this point [A] and put it there [B] it will fit in.
Author: Yes it will fit in, but how much do you have to turn the whole triangle?
Mary: But how many degrees. . .is that one hundred and twenty?
Author: That's right.
Mary: *It's the external angle.*
Author: Yes.
Mary: Ah. . .now how would you define that? How would you put it in simpler terms?
Author: For the children, you mean?
Mary: *For children and for myself.*

Mary was keen to express her insight in simple terms for her notes. She explored the relation between the exterior angle and the symmetry of a figure a little further. She found that a square agreed with her conjecture but that a rectangle did not, because one had to rotate a rectangle by 180 degrees to fit it back in its frame. She also found that her conjecture did not work for an irregular pentagon. Eventually she understood that her rule worked only for regular polygons and expressed its essence unconventionally as follows:

> 'The rotation symmetry is [the total] external angle divided by the number of sides of the regular polygon.'

Next session, Jane was exploring the pattern in the polygon table but was unable to explain it. I suggested that Mary show her. Mary tried to show Jane how polygons could be divided into triangles to illustrate the rule. Mary showed her the triangle and quadrilateral but forgot how to partition the pentagon correctly. Instead of the pentagon in Fig. 7.14, she drew one as in Fig. 7.16 which was divided into five triangles rather than three. I had to intervene to help her with her drawing.

Despite this final small setback, Mary's programming work with polygons had been most successful. From an initial contradiction about the angle properties of a star, Mary had moved on to a more general consideration of

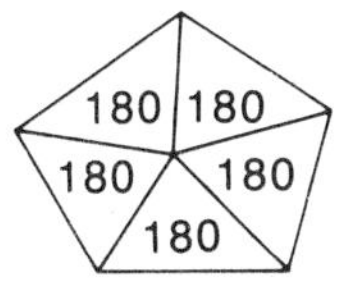

Fig. 7.16 — Wrong division into triangles.

the angle properties of polygons. Then she had discovered a pattern among the angle properties of polygons and she had constructed an explanation for her pattern. Finally she had discovered a relation between the symmetry of a regular polygon and its exterior angle. The value of the programming work was that it stimulated the initial problem and provided an environment in which Mary could test her mathematical insights.

Where Irene had been overwhelmed by the complexities of programming, and where Jane had seen a pattern but had been unable to understand it, Mary had used the programming as a successful jumping-off point for a piece of personal mathematics.

Coordinates

Coordinates were introduced to Mary as a method of positioning her drawings anywhere on the display screen. Her initial reaction was:

> 'I have seen this in maps, but I haven't seen it in maths.'

Like Jane, Mary was not familiar with the turtle's interpretation of negative arguments for FORWARD, BACKWARD, LEFT and RIGHT. So she expected to define different procedures to drive the turtle to different parts of the screen. She had little difficulty in defining a procedure which would take the turtle from the centre of the screen to any point in the top-right hand quadrant. After some prompting from me Mary tried her procedure with its first argument value negative and then with its second argument value negative. She predicted that with two negative arguments, the turtle would be driven into the bottom left-hand quadrant of the screen. She then confirmed her prediction with evident satisfaction.

The value of this work was twofold. The procedure explicitly described how the ordered pair of coordinates was to be interpreted, i.e. so much along and so much up. It also provided her with a means to position her pictures which she used many times thereafter with both positive and negative argument values. This provided a useful, personal application for negative integers.

Later in the term, Mary was observed teaching coordinates in school. The lesson had been inspired by a mathematics programme for schools on television. Mary and Jane listened to the recording of this lesson and Mary explained how her prior experience of coordinates in the programming sessions had helped her:

Author: Did you find it helpful to have done it [coordinates] here?
Mary: Yes it did because the teacher [the children's class teacher] *was going to stop doing the workshop because she found it very, very difficult to follow.*

(a little later in the conversation)

Author: . . .So what I wanted to get at was, did you find that er, the work you'd done on coordinates?
Mary: Did help me, yes u-hu.
Author: So you had a strong idea about what was.
Mary: Yes about what was happening there.
Author: U-hu.
Mary: Otherwise I think I would have been lost myself.

Mary continued by explaining that coming to Logo was very helpful.

Mary: Em, I mean this is what I said, I mean they asked me 'Why do you go there [to Logo], is it voluntary?' and I said 'Yes it is' you know.
Author: Yes, what, you mean here?
Mary: Yes coming here and I said 'I'm really glad that I've learnt quite a lot'.
Author: Mm.
Mary: In fact I sometimes find it is strain coming, you know, like when it is raining and so on. . . .

Mary explained how she had wondered initially what the purpose of the programming work had been, but that now she was much happier because she was tackling mathematical projects related to her school work:

> 'No, I think especially this term, I mean after coming with my problems. Because I thought the first, you know, I thought the first two years when I had been coming [she has overestimated], *I was wondering what is the point, you know. But I think this time it has been really worthwhile.'*

During this term and the next, Mary brought many 'problems' to the Logo sessions. At this session she asked about the convention for the order of the coordinates in a pair:

> 'Now just tell me Ben, I don't know, I have never figured out why you read the bottom number first [i.e. the x coordinate and then the y coordinate].'

I replied that it was just a convention. At a later session, Mary recounted a related difficulty. Her pupils had constructed a scatter graph of their heights (a vertical measure) against their arm-lengths. Mary wanted to know whether the children's heights *had* to be represented on the vertical axis of the graph. She could not decide whether this was merely a matter of convenience or of greater mathematical importance.

Understanding Functions Computationally

This section shows how Mary learned about functions by considering a set of hypothetical procedures. She did not need to run these procedures, but just studied their properties. Mary took the same special mathematics course in the College as Jane. She also found that some of the topics, such as symmetry groups, were a little mystifying:

> '. . .I can understand the simple reflection. . .but when he [the College lecturer] started the complicated stuff, you know, I couldn't understand. And when I asked him, you know. . *.the way when they look at you as if that's easy. . .we are really stupid and ignore us. . .why did you do the course, you know.'*

Mary, like Jane, confused a transformation with the result it produced. She also had trouble filling up group tables and using the information in the tables to solve equations.

Mary had been given a worksheet on groups by the College. She found many of the questions very difficult. Her exploration of one question in particular is now described because it shows how we were able to make use of her prior programming experience. Several procedures were written in the course of this work but they were not run on the computer. It was sufficient to rephrase the question in computational terms to make it more understandable.

The question from the College was to fill in a group table given:

> The set of mappings $\{i,j,k,l\}$ on real numbers defined thus:
> $i{:}x \rightarrow x$
> $j{:}x \rightarrow 1/x$
> $k{:}x \rightarrow -x$
> $l{:}x \rightarrow -1/x$
> together with the 'multiplication' * where $j*k$ means 'j, and then k'
> e.g. $j*k{:}x \rightarrow 1/x \rightarrow -1/x$

Mary did not fully distinguish between the mapping and the variable:

> '. . .so can you say *the value of j is one third, one over x?'*

Mary preferred to work with specific values of x, e.g. 3, rather than with the variable. I showed her how the mappings could be represented as procedures. The name of the procedure was the name of the mapping and the variable x was the argument. I defined the mappings i and j as follows:

```
DEFINE "I "NUMBER
RESULT VALUE "NUMBER
END

DEFINE "J "NUMBER
RESULT DIVIDE 1 VALUE "NUMBER
END
```

Because the Logo implementation only supported integers and not reals, the functions defined by the procedures had different domains and codomains to those defined in the College worksheet. This point was not mentioned to Mary. Since the procedures were not run on the computer the effect of the integer division 'DIVIDE' was not important (it would have rounded down the quotient to the nearest integer).

By the time the first two procedures had been written down, Mary was beginning to get the idea and anticipated the result of procedure 'K' as 'minus the number', and that of 'L' as 'result, subtract, result over, isn't it?'.

```
DEFINE "K "NUMBER
RESULT SUBTRACT 0 VALUE "NUMBER
END

DEFINE "L "NUMBER
RESULT SUBTRACT 0 DIVIDE 1 VALUE "NUMBER
END
```

I used the clumsy form 'SUBTRACT 0. . .' because there was no standard prefix form of MINUS in Logo. I then distinguished between the convention in the worksheet that '$j*k$' meant 'j' before 'k', and Logo's method of parsing commands. In Logo the command:

```
PRINT J K 4
```

meant apply 'J' to the result of applying 'K' to 4, in other words 'K' before 'J'.

In the conversation about the action of the mappings, Mary showed that she had not grasped the idea of composition correctly. She thought that '$j*k$' applied to 3 meant 'do j to 3 and then do k to 3 as well'. Her difficulties were increased because she did not know the identities:

$$1/(-x) = -(1/x)$$

$$1/(1/x) = x$$

Mary considered the composition *'i*l*j'*, but could not see how *'j'* worked on the result of *'i*l'*. So I compared this composition to its equivalent Logo form:

PRINT J L I 3

where 3 was chosen as an arbitrary value for x. I explained how the Logo command would be evaluated and how each procedure would get an input value and pass on a result, see Fig. 7.17. Mary then tried to explain this back

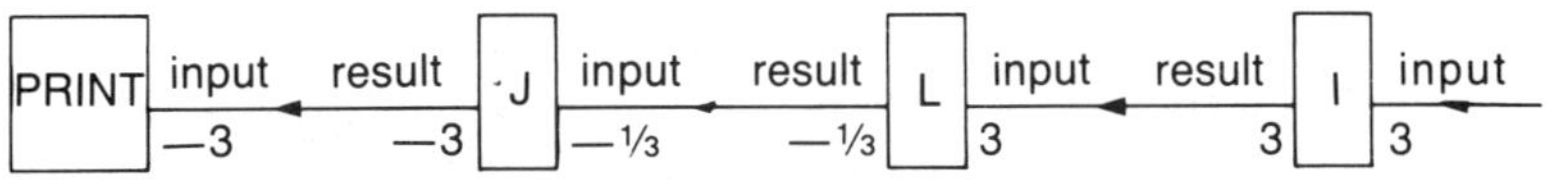

Fig. 7.17 — Result passing.

to me. At first she did not distinguish properly between the mapping and its result:

Mary: . . .first 'I', I mean you are giving a number *so 'I' is three.*
Author: Yes, 'I', well 'I', is a procedure and it's given an input.
Mary: Right.
Author: OK.
Mary: Well 'I', 'I' yeah, the input for 'I' is three.

Mary and the author continued to trace the execution of the command.

Author: And J is told, whatever you're given, you put one over it.
Mary: *Ah, whatever you are given, put one over it,* so that makes. . .right now I have got it.
Author: OK.
Mary: I've got it. See, em, that is *x,* we got as far as that.
Author: Yes, that. . . .
Mary: So the work. . .that was. . *.the work of J is to put one over any number that was given to it.*
Author: Yes.
Mary: Previously.

By the end of the dialogue Mary had formulated a clear personal, and procedural, description of the mapping J which carefully distinguished it from the variable. Using the Logo procedures, I showed her how *'l*j'* was equivalent to *'k'*. Mary again siezed the initiative and asked if she could explain this back to me in her own words. She then considered the procedures which had been specified. Beside each one she wrote a brief note, as she described its action, see Fig. 7.18.

Mapping	Mary's note
I	number
J	1
	number
K	−N
L	−1
	N

Fig. 7.18 — Mary's notation.

Mary's personal notes on the action of the mappings were very similar to the formal description given by the College. The procedural description had served an important function of explicating the original notation. Now that Mary understood the meaning of the mappings, the rather long-winded procedural representation could be abandoned for a more concise notation. The procedural representation had served a temporary purpose. This was not to replace conventional notation permanently but to explicate it.

Mary explained how pleased she was at the end of this part of the session:

> 'Ah, I can see here that Logo is really helpful making. . .I am pleased it has really worked for me.'

Negative integers and subtraction

Mary explored the effect of negative integer arguments values on the procedure for generating Fibonacci series described in Jane's case study. The important distinction between the meaning of the symbol for a negative integer and the symbol for the operation of subtraction was brought into sharp focus when she was puzzled by the behaviour of this procedure. In Logo the operation of subtraction was called by running a prefix procedure, SUBTRACT. The symbol '−' was reserved for negative integers.

```
SPAGHETTI 200 −10
200
−10
190
180
370
550
920
1470
```

Each term was the sum of the previous two, where the two argument values,

200 and −10 were taken as the first two terms. But Mary was perplexed. It seemed to her that the terms were first decreasing in size and then increasing:

> 'I see that, but the thing is I am trying to work out. . .if I give it, eh, a minus thing, *why doesn't it go [on] deducting?*'

Mary explained the 190 in the series as 'two hundred *take away* ten' and the 180 as 'a hundred and ninety *take away* ten'. But she explained the 370 as 'adding'. That is, she interpreted the series as first subtracting and then adding rather than as consistently adding, initially with negative terms. She also noted that the relation between the terms could be expressed in a different way:

> 'You can work it the other way round. You can say three-seventy take away a hundred and eighty [makes 190]', i.e. if
>
> $I(n) = I(n-1) + I(n-2)$ then $I(n) - I(n-1) = I(n-2)$.

Mary did not establish whether this applied to the early terms. She also explained that she really wanted the procedure to deduct 10, then deduct 20, then 30 and so on. I suggested that she write such a procedure. I also suggested that she compare the action of the spiral generating procedure with the Fibonacci series procedure. The former would help to reveal the recursive structure and could be easily modified to produce the series which she wanted (by printing the lengths of the line segments). Mary's comments suggested that she might have been muddling up subtraction and negative integers. For example when describing the procedure she wished to define which deducted increasing multiples of ten, she said:

> '. . .I want, er, SPAGHETTI to give the numbers of negative, I mean the positive numbers *deducting the negative numbers from the result*'

Mary tried the procedure named SPAGHETTI again with two negative argument values:

```
SPAGHETTI −200 −10
−200
−10
−210
−220
−430
−650
```

Mary explained its action:

Mary: In fact that is all right because, er, that is negative and negative, that's all right because negative and negative, it is er. . .*you shouldn't add negative and negative numbers.* So two hundred plus ten that is two hundred and ten. Two hundred and ten plus ten that is two hundred and twenty.

Mary had the idea of modifying the Fibonacci series procedure by replacing its addition by subtraction, i.e. so that it would generate the series:

$$I(n) = I(n-1) - I(n-2)$$

She made the modification and tried out the new procedure, which she named FIBONACCI:

```
FIBONACCI 200 −10
200
−10
210
−220
430
−650
```

I asked her if she could explain why the third term was 210, but she could not. She traced the action of the procedure and then I asked her more specifically:

Author: . . .so from two hundred subtract minus ten [200 − (−10)].
Mary: Aha.
Author: Now do you know what happens if you subtract a negative integer?
Mary: From a positive.
Author: Yes.
Mary: *I don't know with Logo but it becomes one-ninety.*

I pointed out this mistake and explained the difference between:

```
SUBTRACT 200 10
SUBTRACT 200 −10
```

I also pointed out that '−' conventionally had two meanings and that the Logo expressions above could be written:

$$200 - 10$$
$$200 - (-10)$$

Mary then remembered a 'rule',

'oh so that minus and minus changes into plus'.

Mary decided to try out her new procedure FIBONACCI with pairs of positive numbers, e.g.

FIBONACCI 2 20
2
20
−18
38
−56
94
−150

She spent some time trying to understand how the procedure produced the given results. Part of her effort was devoted to understanding argument binding in the recursion. But she also spent time trying to explain to me why the terms of the series alternated in sign. This demanded that she subtract negative integers from positive. Thus she explained the expression 'SUBTRACT 18 −16' which had been evaluated during the run of her procedure as:

'Take away, take away. . .so it becomes positive [34]'

having initially suggested '2' as the answer.

The value of this programming work was that it provided an interesting problem in which addition and subtraction of integers played a part. The programming language distinguished between the operation of subtraction and the symbol for a negative integer. It is unclear whether Mary's phrase, 'take away, take away. . .so it becomes positive' is evidence that she had muddled this distinction or whether it was just a convenient rule of thumb for dealing with such subtractions.

When Jane's work with the Fibonacci series is compared to Mary's we find that Jane spent nearly all of her time trying to understand how the recursion worked. In contrast, Mary investigated the recursion, the properties of the series and integer operations.

Mathematical curiosity

Mary was more curious about arithmetic than either Jane or Irene. For example, she asked me why the algorithms for addition, subtraction and multiplication all started on the right with the least significant digit, whereas the algorithm for division started on the left with the most significant digit. She had also been shown a quick method for long-division by a teacher. He showed her how she could change a division by 16 into two successive divisions by 4, so long as she carried out the appropriate computation on the two partial remainders. The example below divides by 24.

Division by 24 in one stage:
24)397
—
16 Remainder 13

Division by 24 in two stages (24 = 6*4):
6)397
—
4)66 Remainder 1
—
16 Remainder 2

The particular example Mary cited was division by 16 (changing ounces to pounds). But she could not remember how to deal with the remainders. She realized that she could not just add them together since it gave an incorrect remainder in some cases. She asserted, incorrectly, that for division by square numbers such as 16 she could find the total remainder by adding the first partial remainder to the first factor but worried:

'then I don't see any logic in it.'

Jane, who was present, remembered doing similar calculations but was not able to help. Mary explained that she and her sister had tried to find out how to do these calculations but had been unsuccessful.

At the next session, Mary was given a procedure to run so that she could try to solve this remainder problem for herself. The procedure, named BLOCKS, took two numerical arguments. The first value was divided by the second and the result of the divisions was illustrated by printing sets of 'X's and 'R's as follows:

```
BLOCKS 13 3
XXX XXX XXX XXX R

BLOCKS 21 6
XXXXXX XXXXXX XXXXXX RRR
```

Mary tried out this procedure on a number of divisions such as:

```
BLOCKS 27 6
XXXXXX XXXXXX XXXXXX XXXXXX RRR
```

This illustrated that 27/6 was 4 remainder 3. Mary then compared this division by 6 with successive division, first by 3 and then by 2.

BLOCKS 27 3
XXX XXX XXX XXX XXX XXX XXX XXX XXX

then having counted the sets of 'XXX's

BLOCKS 9 2
XX XX XX XX R

The problem was to relate the single remainder 'R' in this last command to the remainder 'RRR' when dividing 27 by 6.

Once Mary thought she had solved the problem she wrote out the division on paper using a representation scheme similar to the 'X's and 'R's of the procedure. When she was satisfied that she understood the mechanism, she explained it to me using her pencil and paper diagram.

Mary also explained the more complex comparison of '79/12 = 6 R 7' with '79/2/3/2' which she had tried earlier using the procedure BLOCKS.

Mary: You are dividing it [79] by 2, you are getting thirty-nine blocks.
Author: Yes.

(**)(**)(**). . .(**)(**)(**)*

Mary: Of each block containing two units.
Author: Yes.
Mary: And then getting one unit remainder.
Author: Yes.
Mary: Then you are dividing those thirty-nine blocks of two units in each block.
Author: Yes.
Mary: By three.
Author: Yes.

[(**)(**)(**)] [(**)(**)(**)] [(**)(**)(**)]
[(**)(**)(**)]. . .[(**)(**)(**)]*

Mary: So you are getting thirteen blocks.
Author: Yes.
Mary: Right.
Author: Exactly.
Mary: OK and then those thirteen blocks have got six units.
Author: Yes right.
Mary: In them, you are dividing them by two.
Author: Yes.

{[(**)(**)(**)][(**)(**)(**)]}. . .{[(**)(**)(**)][(**)(**)(**)]}
[(**)(**)(**)]*

Mary: So you are getting six whole blocks and then you are having a block left over which has got six units in it.
Author: Yes.
Mary: So six units and first one unit here, your remainder is seven units.

I could have explained the solution to Mary without use of a procedure. But by running the procedure and by attempting to relate its effects to the problem, Mary had the satisfaction of solving the problem for herself. The action of the procedure suggested a useful way of looking at the division. The procedure made one facet of the division process more explicit, not through the body of its code, but through its effects. Mary did not know how the procedure was defined. We may contrast this example with her work on mappings described in the last section. In both cases a procedure provided an explicit, but temporary descriptive system for a process. In the case of the mappings it was the text of the procedure which provided the explicitness. In this latter case it was the effects of the procedure. In each case Mary invented her own simpler notation to deal with the problem, once she had grasped its essentials.

7.3 SUMMARY

Mary's and Irene's experience of programming is summarized using the same framework which was applied to Jane's work.

Rigour and explicitness

From the description of Mary's work it is clear that no attempt was made to teach her to derive theorems by rigorous argument from clearly defined axioms. The programming work undertaken did not address the issue of 'rigour' in this sense. What it did do, however, was to illustrate the idea of 'explicitness' and the value of unambiguity in a formal language. Such 'proofs' as were constructed, e.g. about the interior angle properties of polygons, were not rigorous but highly plausible arguments.

While Mary was sometimes frustrated by Logo's literal interpretation of her commands (as against 'understanding what she meant to say'), she was not overwhelmed by the details of programming in the way that Irene had been. This is not to say that Mary did not spend much of her time considering programming issues rather than mathematical issues. But she was able to cope with the programming and attack many problems at a more abstract, mathematical level.

Like Jane, Mary saw the value of forming explicit descriptions. In answer to the question, 'What do you think you have learned since doing LOGO?', from the third questionnaire, she replied:

> 'A [sic] break a problem in steps [sic]. To get thing [sic] clear in my mind and predicting the difficulties it might cause if my explanation [sic] is not clear.'

A good example of the way that consideration of Logo programming made a process more explicit was Mary's work on functions. The notation used by the College was concise and elegant.

$$k{:}x \rightarrow -x$$

But Mary's unfamiliarity with it meant that she did not fully appreciate the different status of the names 'k' and 'x'. Mary's greater familiarity with programming notation and with the action of procedures made the more verbose, procedural description a better vehicle for the idea. Once she had understood the idea she reverted to a notation very close to that of the College. The procedures, which were never run, played a temporary but important role in making the idea of a mapping explicit.

On other occasions programming served to focus her attention on relevant issues. Thus work which started by drawing triangles and squares finished with an examination of the relation between the exterior angle of a regular polygon and its symmetry. The programming specification of regular polygons as sequences of 'FORWARD n, LEFT m' made such a progression possible by reducing the descriptions of the polygons to their bare essentials.

Irene found programming harder to learn than either Jane or Mary. She became much more frustrated by programming setbacks. The constraints imposed by the programming often seemed pointless to her, espeically when she matched the simplicity of the picture 'products' of a session against the complexity of the process of programming the drawing. She planned her procedures less than either Jane or Mary and was less interested in understanding why a procedure worked.

Active exploration

Mary's case study has shown how the programming work acted as a catalyst for mathematical exploration. Where Irene had struggled with programming, Mary was often able to attack a problem on a more abstract, mathematical level. Mary's search for explanations was more aggressive than Jane's, so she was more persistent in looking for the mathematics underlying some of the programming tasks.

Several examples were given where Mary was stimulated to ask mathematical questions arising from the programming work. For instance, her work on the angle properties of polygons was grounded in her early turtle drawing. Her analysis of the subtraction of positive and negative integers was pursued in her attempts to understand the printed output of a series-generating procedure. Her solution to the problem of the comparison between a single division and successive division by factors was prompted by her examinations of the effects of running a procedure. Her work on coordinates and negative integers had the immediate benefit of enabling her to position pictures easily on the display screen.

Few instances were found where Irene asked mathematical questions about the programs or drawings she produced. Her behaviour contrasted very strongly with that of Mary, in this respect. Irene repeatedly grappled with the problem of understanding the way the turtle executed rotations but never seemed to sort it out fully. Most of her problem-solving was conducted more by trial and error methods than through an analysis of the problem. This had the effect that, when a solution was achieved, she had not benefited mathematically from the experience.

Key concepts

Mary studied much the same range of mathematical concepts as Jane. Again the main emphasis was on geometry, but useful work was done on the topic of functions and variables. This showed how even a 'pencil and paper' consideration of Logo procedures and their execution could help clarify such issues as composition and the distinction between functions and variables. In contrast, Jane had addressed these issues by matching a transformations to running a procedure, whose definition she did not know.

Irene's grasp of angle was still poor at the end of her studies. The partition/quotition distinction was only partially explored through programming.

Problem solving

Some of Mary's answers to the third questionnaire suggest that she had understood the value of problem decomposition. In answer to the question. 'What was the most useful thing in learning Logo?'. Mary replied:

> '*Learning to work a problem step by step*. When I could see the use of Logo in connection with maths and how to solve problems using Logo's procedure. Seeing the results of your procedures.'

Another question, 'Since doing Logo, have you found that the way you approach a problem is different? If so, in what way?' Mary responded:

> 'Take a problem *step by step* and try and explain in different methods, if one method fails.'

Mary's comments during the programming sessions show that she valued the opportunity to solve problems presented by the programming work. At the end of one session, Mary explained that what she had enjoyed was:

> 'Now especially this achievement that was good, you know, em, *where I could find my own mistake and where I could correct it.*'

Irene made only a few comments about her problem-solving strategy. On one occasion she explained that she was constructing a solution 'little by little'. Irene expressed general negative opinions about her own ability. These were not sufficiently specific to suggest remedial action on her part:

> 'My brain isn't logical, that's why I don't get on with this.'
>
> 'My brain needs a good shaking just now.'

When asked to make a comparison between her experience of programming and a child's experience of mathematics (the second questionnaire), she concentrated on her feelings about the subject rather than on the methods of tackling it:

> '. . .the first analogy is very much how I feel about Logo. Logo like maths is/was very alien to me.'

Thinking about learning

Like Jane, Mary used her experience of programming to think about problem-solving and learning. Mary explained that there were times when it was better to be told the answers but that:

> '. . .to puzzle it out yourself and then getting the hint here and there. . .eh, helps you in fact. . .I think it leads you into. . .if you come across something else then you can link up the other past experience into it.'

Later that term, Mary had an insight into the reciprocal nature of teaching when she was trying to work out the meaning of an error message from Logo:

Mary: I can see it now.
Author: You can see it now.
Mary: Do you know what I can see? When the child doesn't understand the teacher and the teacher doesn't understand the child, the frustration starts.

The majority of Irene's comments about programming referred to her emotional reactions to it rather than to any insight into her own problem-solving behaviour. Although Irene said that programming made her 'really think', the context suggested that on the whole this was an unpleasant experience rather than an enlightening one. She did not look on her mistakes as constructively as either Jane or Mary and saw them as frustrating setbacks.

Attitude to mathematics

Mary was more curious about mathematics than either Jane or Irene. As well as bringing her mathematical difficulties from the classroom or from College, she also brought bits of mathematics which had intrigued her. An example was the problem of comparing a long division with successive division by smaller factors. Her initial attitude to mathematics was different to that of Jane and Irene. They all, rightly, knew that they did not understand much mathematics. But Mary seemed to believe that she could learn to understand and she did not invest the subject with Jane's pessimism or with Irene's feelings of incompetence.

The programming work stimulated Mary's mathematical curiosity, as in the case of the six-pointed star, and more importantly, helped her to make personal discoveries. Thus, after she had confirmed her conjecture about the pattern in the table for the angle properties of polygons, she said happily:

> 'I have discovered something new, yes.'

Mary consistently took a more active part in conversation with me than either Jane or Irene. She would demand that I re-explain some point or she would try to explain the point back to me. She liked to put things 'in her own words' and to make notes. She did not have the same self-consciousness that Irene had and did not mind whether I watched her or not, so long as I was available to give help when she wanted it.

At the beginning of the study Mary was rather deferential towards me and apologized for asking questions. But as time went on she felt at ease and made several remarks about the style of the Logo sessions. Thus at a session to which she brought no mathematical difficulties, she announced, 'No confessions today'. At another session she said that recounting her mathematical difficulties was like 'telling sins in confession'.

When working with Jane on integers, Mary commented, with some amusement, that I had been using 'guided discovery'. These comments reveal a rather less tense atmosphere than when I was with Irene.

Irene gave no indications that her attitude to mathematics had changed. Irene made explicit comparisons between the unpleasant feelings she experienced while programming and those feelings she remembered from mathematics.

> '. . .my mind goes exactly the same way as it does in maths, exactly the same.'

Although Irene did eventually solve some of the programming problems, the experience seemed to remind her of earlier mathematical failures rather than to lead her to expect future mathematical successes.

Irene made few references to the way she was taught in the College of Education. Unlike Jane and Mary, Irene was not taking any extra mathematics course. Irene did not like being watched while she programmed and preferred either to seek help herself from me or for me to provide small amounts of specific advice. Unlike Jane she did not like the situation where she was attempting to solve a problem to which I knew the answer, since it made her feel stupid. She also found that teaching practice lesson observation made her nervous and very much more self-conscious about using the 'proper' mathematical vocabulary. She also felt 'inhibited' by the one-to-one discussion with me about the lesson recording. For her last lesson observation, I gave Irene the tape first, so that she could listen to it and come prepared with her own comments. This seemed to help her a little though most of her comments were about her speech mannerisms than about the mathematics she was trying to teach.

Disadvantages of programming

Much of Mary's work was concerned with programming rather than with mathematics. She made several references to the fact that most of the initial programming sessions had scant mathematical relevance but were concerned entirely to teach programming. In the third questionnaire, answering the question, 'What was the worst part of Logo work?' she answered:

> '(a) When I could not see immediate value of Logo work with the teaching of maths (2nd year) [she joined the study in her second Diploma year].
> (b) When procedure [sic] made by me would not work out due to not enough inputs or language that Logo did not understand.'

Earlier she had made the same point during the polygon work when it became clear that something of mathematical benefit was happening:

Mary: It's starting to get more [inaudible word] now, it's getting into real mathematical things.
Author: It is, isn't it? Do you think this is a reasonable way to explore mathematics?
Mary: Ehm, I think so yes, u-hu. Because at first I couldn't see the real meaning behind the. . .I mean what was I doing with Logo all the time. . .'

Mary contrasted the early work, predominantly concerned with programming, with the later work concerned with her analysis of her own mathematical difficulties conducted through programming.

> 'No I think especially this term, I mean after coming with my problems, because I thought the first, you know, the first two years [she has overestimated] when I had been coming, I was wondering what is the point. But I think this time it has been really worthwhile.'

Mary, like Jane, also attempted a number of topics, such as drawing fraction pie-charts and representing vectors as lines on the graph-plotter in which the mathematics was studied at the wrong level of representation. Mary's experience was similar to Jane's in this respect and it was not re-described in her case study.

Much of Irene's work was concerned with programming rather than with mathematics. The difficulties she had with the programming ovrshadowed much of her work. Very often she saw her task as the production of a particular picture, a 'house' or a 'flower' rather than a search for underlying geometric principles.

In more open-ended problems, such as exploring the action of a given procedure by supplying it with different argument values, Irene seemed content to supply a few values only and she formulated few questions about the action of the procedure. Her work with a spiral procedure did cause her some surprises, for example the production of a polygon when she supplied an increment of zero. But she only seemed to seek causes for these surprises when asked to by me. She seemed to have little mathematical curiosity and never experienced the obvious delight of moments of insight which Mary enjoyed.

8

Conclusions

I have tried to give a detailed account of how three students learned, and failed to learn, mathematics through their programming activity. The three students were typical of the 15 students who took part in the study. This has shown that programming as a method of learning mathematics has a number of benefits for such students. Programming encouraged a constructive view of mistakes and difficulties. The emphasis in the programming classroom was on understanding why the programs worked or did not work. There was usually little shame associated with a computer program which did not run correctly the first time. The need to make commands to the computer explicit forced the students to acknowledge and explore the topic which they did not understand.

A number of key concepts were given concrete illustration and explored by the students. Concepts with a dynamic element were especially well-suited to this method of exploration, e.g. angle as rotation, functions as transformations and the transformation/state distinction. The students were able to explore a domain by observing how the primitives behaved and interacted. Some success was achieved in explaining mathematical concepts by reference to the code of hypothetical procedures and commands. Here I and the students planned and hand-traced code but did not run it on the machine.

Students were able to pose and solve many problems in the domains defined by the primitives. Students gained a great deal of satisfaction by solving such problems. Mathematics was presented as an exploratory activity in which personal discoveries were possible. Turtle Geometry was used extensively. Students used this domain to ask mathematically interesting questions and to experience 'aha' moments of insight. Students found that they were capable of solving problems and usually enjoyed writing and debugging simple programs to draw pictures.

But there were difficulties in implementing a scheme based on programming. The teaching of the prerequisite programming skills was overemphasized so that it distorted the overall mathematical objectives of the work. At first the students did not see the relevance of their programming to their future work as teachers. The programming should have been been introduced in a mathematical context and the specific links between programming and mathematical ideas pointed out from the start.

Initially some students were asked to represent mathematical processes in the notation of computer programs. But later in this study it was found that the major difficulty of the students was not their knowledge of how mathematical processes worked but their lack of understanding of why they worked and what they meant. For example, some students knew how to add, subtract, multiply and divide natural and rational numbers. But they did not understand the applications of these operations and they did not have explanations and illustrations for them which they could use in the classroom. Rewriting these processes as computer programs was a pointless activity when the work concentrated on the mechanics of the process rather than on its meanings.

Logo had a small range of data-types. This meant that it was necessary to write new primitives to study many mathematical topics. This was often an inappropriate task for the students to undertake because the majority of the resultant problems were concerned with programming rather than with mathematical issues. It proved more effective to provide students with these new primitives. This freed them to concentrate on the mathematical properties of the primitives.

Although the students learned how to write simple programs to control the turtle, they found complex programs involving symbol manipulation hard to plan and debug. This meant that many existing projects based on programming were too complex (e.g. Lukas *et al.*, 1971). The students had limited time available in a crowded Diploma timetable. They did not see the value of investing large amounts of time in learning programming. One student even found programming just as unpleasant and difficult as the mathematics it was supposed to elucidate.

There was a major difficulty in approaching mathematics through programming using drawing devices. The planning and debugging strategies observed among the students were their response to the ready availability of the computer and betrayed their preoccupation with the drawings as 'products' rather than with planning and analysis. Easy access to the computer encouraged them to plan and debug their programs at the console. Students were disinclined to analyse their programs once they appeared to work to see whether a more efficient or elegant solution could be formulated. Further attempts to teach mathematics through programming will have to guard against these tendencies and stress the importance of careful analysis, away from the console, both before and after a program is written.

It was difficult to design programming projects which successfully confronted the students with their individual mathematical difficulties. Projects were unsuccessful when the mathematics was overwhelmed by the programming. For instance, asking students to write programs to *draw* representations of mathematical structures and processes was usually ineffective because the students concentrated on the *visual* properties of the representation rather than on the underlying mathematical properties. Again, programs which manipulated symbols representing objects were cumbersome and it seemed better (in hindsight) to ask the student to manipulate real objects. Representing integers as turtle movements was successful, however, because of the simple mapping between the two and because the integer operations could be investigated without writing complex programs. Ideas for teaching a number of other topics have been described, including fraction operations and symmetry tansformations.

It was important to provide students with good teaching as well as good projects. Learning mathematics through programming did not mean that there was no need for help from a human teacher. The case-studies gave numerous examples of where I lost opportunities to exploit a situation mathematically. The case-studies also showed that it was necessary for the links between the programming ideas and the students' existing mathematical knowledge to be made explicit. Communication between the student and the computer had to be supplemented by communication either with a teacher or with other students. Students should explain their programs in English to someone as well as writing them in Logo.

The experiment described in Chapters 5–7 was, necessarily, conducted with only a small group of students. This group was not a representative sample of student teachers and they were taught outside their College of Education in an University Research Laboratory. I acted as the students' tutor, as an experimenter and as the evaluator of the work. Later research tried to broaden the scope of the study by working with larger and more representative samples of students and by conducting the work more within the normal course framework of a College or Department of Education.

The mathematical work undertaken by the students was largely determined by their mathematical difficulties. These difficulties were revealed in an ad hoc manner by classroom observation, by discussion and, to a limited extent, by the mathematics test which was administered. In later work this was improved in a number of ways.

In this experiment, the teaching of programming was initially overemphasized and was taught with insufficient reference to mathematics. This was improved by changing the role of programming in the mathematics course. The emphasis was shifted towards mathematics and away from programming. Programming was introduced in those parts of the course which merited it, and the programming projects tried not to distract attention from the mathematics. New sets of primitives were developed which, like those for Turtle Geometry, produced some mathematically interesting behaviour without the need for complex programming knowledge on the part of the student. Details of the suubsequent work with student teachers can be found in du Boulay and Howe (1981a, 1981b, 1982) and in du Boulay, Howe and Johnson (1982).

REFERENCES FOR PART II

du Boulay, J. B. H. (1978) 'Learning primary mathematics though computer programming', Ph.D. Thesis, University of Edinburgh.

du Boulay, J. B. H. (1980) 'Teaching teachers mathematics through programming.' *Int. J. Math. Educ. Sci. Technol.*, **11**, 3, 347–360.

du Boulay, J. B. H. and Howe, J. A. M. (1981a) 'Student teachers attitudes to mathematics: differential effects of a computer based course' . *Computers in Education*, R. Lewis and D. Tagg (eds.), Amsterdam; North Holland, pp. 707–713.

du Boulay, J. B. H. and Howe, J. A. M. (1981b) 'Re-learning mathematics through LOGO; helping student teachers who don't understand mathematics.' *Microcomputers in Secondary Education: Issues and Techniques*, J. A. M. Howe and P. M. Ross, (eds.), London: Kogan Page, pp. 69–81.

du Boulay, J. B. H. and Howe, J. A. M. (1982) LOGO building blocks: Student teachers using computer-based mathematics apparatus.' *Computers & Education*, **6**, 1, 93–98.

du Boulay, J. B. H., Howe, J. A. M. and Johnson, K. R. (1982) 'Making programs vs running programs: microcomputers in the mathematical education of teachers.' *Involving Micros in Education*, R. Lewis and E. D. Tagg (eds.), Amsterdam: North Holland, pp. 119–123.

Feurzeig, W., Papert, S., Bloom, M., Grant, R. and Solomon, C. (1969) 'Programming-languages as a conceptual framework for teaching mathematics.' *Report no.* 1889,Bolt, Beranek and Newman, Cambridge, Massachusetts.

Lukas, G., *et al.*, (1971) 'LOGO teaching sequences.' Vol. 3, *Report No.* 2165, Bolt, Beranek and Newman, Cambridge, Massachusetts.

Papert, S. (1973) 'Uses of technology to enhance education.' Proposal to the National Science Foundation, Artificial Intelligence Laboratory, Massachusetts Institute of Technology, Cambridge, Massachusetts.

Part III

Using Logo with very young children

Martin Hughes and **Hamish Macleod**

This part of the book describes a Logo project carried out during 1983–84 in Craigmillar Primary School, Edinburgh. The project is distinguished from many other Logo projects by two main characteristics. First, the children involved were relatively young — at the start of the project their ages ranged from $5\frac{1}{2}$ years to $6\frac{1}{2}$ years. Secondly, the project took place in a severely deprived area on the outskirts of Edinburgh, characterized by many social problems such as bad housing and severe unemployment. Despite, or possibly because of, these two features, the project was a considerable success: when using Logo, the children demonstrated many of the characteristics noted in other Logo projects, such as high levels of absorption, collaborative problem-solving and mathematical discussion. In addition, the children made statistically significant gains on a standard abilities test. The project thus demonstrated that Logo has an important role to play in the education of very young children.

ACKNOWLEDGEMENTS

The research was supported by grants from the Nuffield Foundation and the Scottish Education Department, and we are extremely grateful for their support. We would also like to thank Cathie Potts for her major role in the project, Ann Brackenridge and Jean Fife for their help in producing the manuscript, and the staff and children of Craigmillar Primary School, without whom the project could not have taken place.

9

Why Logo for very young children?

One important feature of the Craigmillar Logo Project was that it was carried out with very young children, whose ages at the start of the project ranged from $5\frac{1}{2}$ years to $6\frac{1}{2}$ years. In this respect it differed from the great majority of publicly reported Logo projects which have generally used older children. In Britain, for example, most of the recent projects have involved children aged 8–12 years (e.g. Noss, 1983; Hoyles, Sutherland and Evans, 1985; Finlayson, 1984) and a similar picture holds for other countries. There are some notable exceptions to this, such as an early study by Perlman (1974), more recent accounts by Vaidya (1984), Berdonneau (1985) and Metz (1985), and the classic case-study by Lawler (1985); but on the whole Logo projects involving younger children are relatively rare.

At first sight this concentration on older children is not surprising. Many of the ideas involved in Logo, such as that of a variable, or recursion, would appear to be difficult for young children to grasp. Even the basic commands of Turtle Graphics, such as 'FORWARD 100', or 'RIGHT 90', involve not only relatively large numbers but also ideas about angle which young children may not yet have encountered. Moreover, the idea of introducing young children to programming computers may seem quite alien to many of their teachers. Why then did we want to introduce Logo to children as young as 5 and 6 years?

The original idea for the project in fact came out of research which one of us had previously carried out in the area of early mathematics (Hughes, 1986). This research showed that even very young children have a good concrete grasp of number, for example in situations involving the addition or subtraction of small numbers of bricks. The basic problem facing these children when learning mathematics in school is one of linking up their concrete skills to the more abstract or formal language of arithmetic. What is needed is that written numerals be introduced to them in such a way that the children can see very clearly what function the numerals are serving. The use of simplified Logo commands like 'F2' (meaning 'forward 2 units') and 'R3' ('turn right 3 units') to control the Turtle seemed likely to provide the children with just this kind of experience. Not only would it teach children about mathematical symbols, but it would also show that these symbols could be useful and powerful — that mathematics could help them achieve goals they had set for themselves.

This argument for using Logo to help children in the first stages of mathematics is very close to one of the arguments put forward by Seymour Papert in his seminal book *Mindstorms*: *Children*, *Computers and Powerful Ideas* (1980). Papert argues that the Turtle provides an ideal bridge (or, as he puts it, 'transitional object') between the abstract world of mathematics and the concrete world of reality. By writing instructions to control the Turtle, children are required to use mathematics in a context where they can see the immediate point of what they are doing. They are thus, according to Papert, likely to grow up without the fear and suspicion of mathematics that characterizes so many of us today.

Another argument that Papert puts forward is that Logo teaches children an approach to planning and problem-solving that will generalize to other areas of learning. In order to produce a drawing or pattern, children must first plan what they want to do, and then break the problem down into an ordered sequence of simpler elements. This must then be expressed in the appropriate mathematical symbols. Next, they must put their program into operation, noting whether the Turtle does what they wanted it to do. If it does not, they must then start a process of 'debugging', or checking the program for mistakes. This process, argues Papert, not only teaches children a healthy approach to solving problems, it also makes them see mistakes as futher problems to be overcome.

Papert also points out that Logo embodies an approach to computers which is very different from that of most other forms of computer-aided learning. He contrasts Logo with 'drill-and-practice' programs, in which the child mechanically works through problems set by the computer, arguing strongly for the open-ended discovery-learning which Logo encourages. 'Let the child control the computer, not the computer control the child' is Papert's message.

It seemed to use that, if Logo could help children in all of these ways, then it should be made available in some form or other to children of any age. Moreover, the principles embodied in Logo (those of open-ended discovery-learning and creative problem-solving) are highly valued in early education in Britain and elsewhere. We therefore set out to see whether a simplified form of Logo could be successfully used with relatively young children.

10

Simplifying Logo for very young children

The equipment used in the Craigmillar Logo Project consisted of three main items: a BBC 'B' microcomputer, a Turtle and a Concept Keyboard. These items, together with the VDU (visual display unit or screen), disc-drive and Turtle control box, were positioned on a specially constructed trolley.

THE BBC MICRO

The heart of the system is the microcomputer itself. Our early investigations were carried out using an Apple II computer, but the machine finally selected for this project was the BBC Micro. There were two main reasons for this choice. First, the BBC Micro is the most widely used educational computer in Britain, and we wanted to work with the machine with which most teachers would be familiar. Secondly, special interfaces are needed to connect the computer to the Turtle and the Concept Keyboard, and these interfaces are standard on the BBC Micro.

One disadvantage of using the BBC Micro was that at the beginning of the project no full version of Logo was available for this machine. While this was not strictly relevant to our immediate concern of developing a simplified version of Logo, it meant that there was nothing for the children to progress to after mastering our simple verision. It also meant that our simplified Logo had to be written in BASIC, a somewhat unsatisfactory state of affairs.

THE EDINBURGH TURTLE

The second item of equipment was the Edinburgh Turtle, originally designed in the Department of Artificial Intelligence at the University of Edinburgh. The Turtle is essentially a floor-crawling robot, which can be controlled by instructions from the computer. It consists of a clear perspex dome containing three motors and a loudspeaker (to enable the Turtle to 'hoot' — being the Edinburgh Turtle this means playing the first few bars of 'Scotland the Brave'!). There is one motor for each of the Turtle's two wheels. If the motors turn in the same direction the Turtle moves forwards or backwards, while if they turn in opposite directions the Turtle pivots about its centre. The third motor is used to raise and lower a pen which can be mounted in the Turtle's centre of rotation. When the pen is lowered, the Turtle draws a line as it moves.

For most of the project the children worked with a Floor Turtle. However, towards the end of the project they progressed to controlling a simulated Turtle on the VDU. This 'Screen Turtle' was in fact a small triangle which could be moved around the screen by using the same commands as for the Floor Turtle. The Screen Turtle could also leave a trail behind it, thus producing a pattern or picture on the screen.

THE CONCEPT KEYBOARD

For many people, and particularly for young children, one of the main stumbling blocks in the way of a painless introduction to computers is the keyboard. Programs can be made very simple, so that the user is required to press only a few keys, yet these still have to be picked out of the whole QWERTY array. What is needed, particularly for a young child, is an input device which presents the user with only those keys which are needed at the time. In other words, an appropriate keyboard would embody, in a clear and obvious way, only those concepts which the program uses. One way of accomplishing this is to use a Concept Keyboard. This is a flat panel with a large number of very small touch-sensitive areas, which essentially allows the teacher or researcher to design their own keyboard. Having done so, and programmed the areas of the panel accordingly, the functional areas can be labelled appropriately. In the course of the project, the children were introduced to a sequence of increasingly complex keyboard arrangements as their proficiency with Logo grew (see Figs. 12.1, 12.5 and 12.8).

TROLLEY

This consisted essentially of a low cupboard on wheels. It contained within it all the equipment apart from the Concept Keyboard and the VDU, which sat on top. Fig. 10.1 shows the set-up in use.

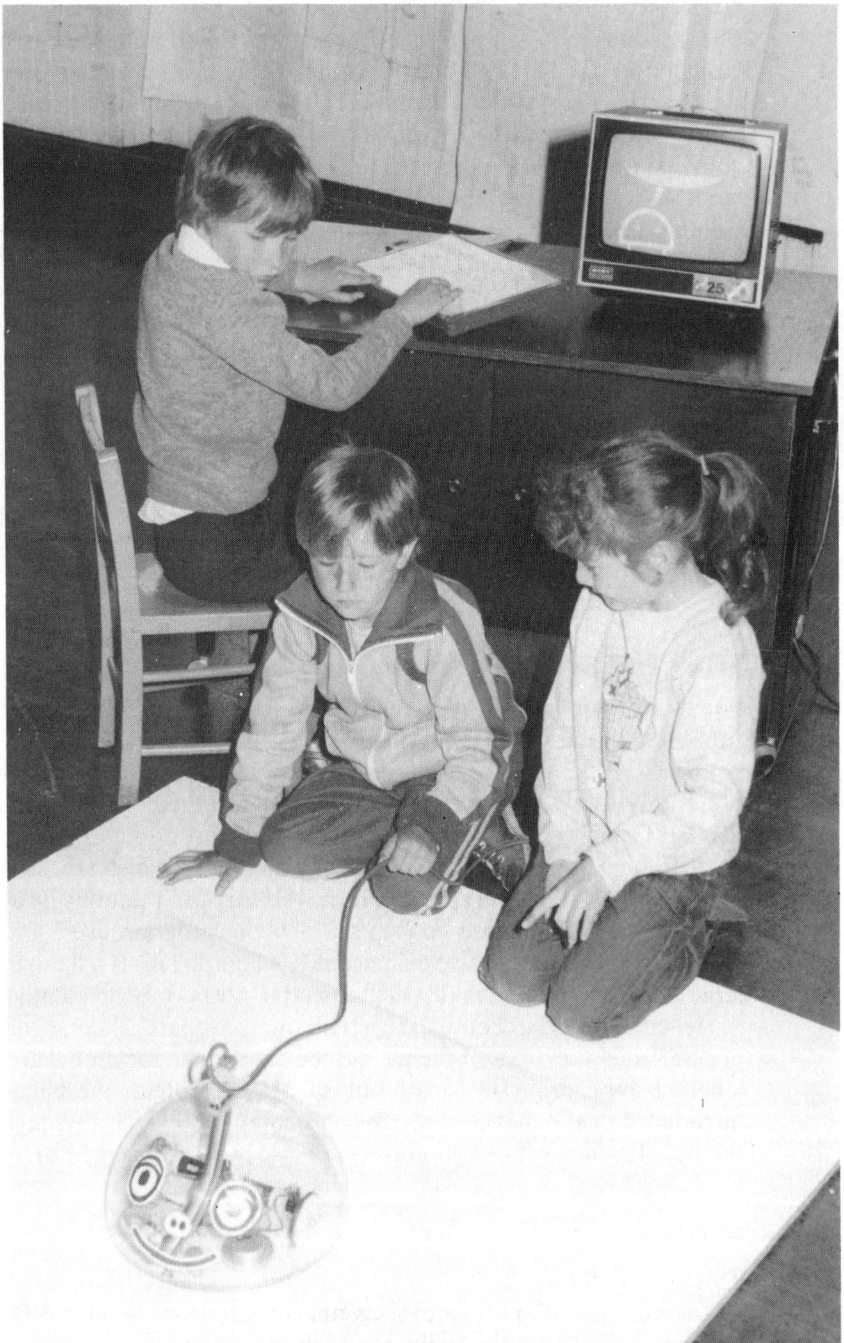

Fig. 10.1 — Set-up in use.

11

Preliminary work with preschool children

Before we started work in Craigmillar Primary School we spent several weeks developing and trying out software for linking the Turtle, the Concept Keyboard, and the BBC Micro. This work was carried out with a group of 5 preschool children from the Psychology Department Nursery at Edinburgh University. Our main priority during this period was developing software, and so the children were sometimes moved on to the next stage before they had exhausted the possibilities of the previous one. Nervertheless their reactions to the Turtle set-up were interesting and worth reporting.

The group consisted of 3 girls and 2 boys. Their ages ranged from 4 years 2 months to 5 years 2 months, and they were all in their last term at Nursery before starting school. They contrasted sharply with the Craigmillar children, being from middle-class homes and manifesting an above average competence in basic numeracy and literacy. At the time they were introduced to the Turtle, all except one child could count accurately up to 10, and they could all recognize the numerals 1–5. All but one could recognize the letters 'F', 'B', 'L', 'R' and 'H' used on the Concept Keyboard. At the start of our work none of the children had a computer at home although, in the course of our work, the parents of one child became so impressed by her interest in computers that they decided to buy one.

The children worked with the Turtle in a separate room away from the main Nursery. They usually came out in pairs, although sometimes they came alone. The children responded enthusiastically to the Turtle and were almost always eager to come out for a session. Once out, they sustained their interest in the Turtle for a substantial period of time: the sessions usually lasted between 20 and 30 minutes and were more often ended by us than by them. The children enjoyed moving the Turtle along paths and roads which they constructed out of bricks. They also enjoyed making patterns on the screen with the Screen Turtle. However, they found it harder to draw pictures with the Turtle's pen on the floor.

In a typical session, two of the girls — Corrie and Susannah — were using a prototype version of the SCREEN program (for more details of SCREEN see p. 183). This allowed them to control the movements of the Screen Turtle by using a simple set of commands. Thus pressing 'F' followed by '1' would send the Turtle forward the smallest distance allowed for, whereas 'F 9' would send it nine times as far. The program also made use of four prepared shapes: if the children pressed the areas of the keyboard labelled with a square, then the Screen Turtle would draw a square — and similarly for a triangle, star and hexagon (called a 'round' by the children).

Susannah had during a previous session discovered that she could make a 'flag' on the screen (see Fig. 11.1(a)) by pressing the following keys:

1. F 5 2. 'square'

One of us suggested she drew a lollipop, and drew one himself on paper to

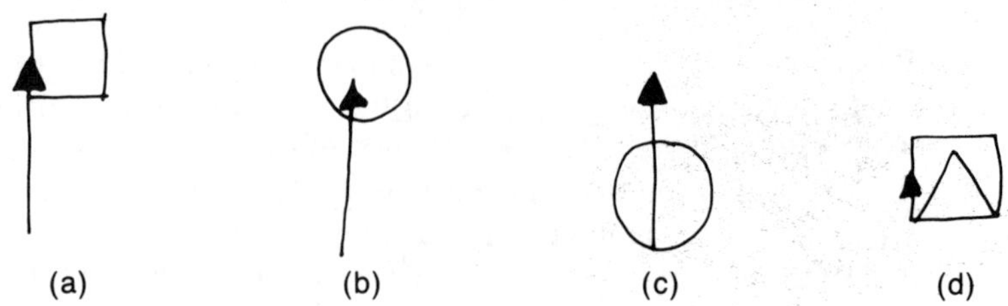

Fig. 11.1 — Pictures produced by Nursery children. (a) Flag. (b) and (c) Lollipop. (d) 'Silly shape'.

demonstrate what he meant. Susannah did not immediately see how she could do this, but Corrie suggested they try 'a line and then a round'. So they tried:

1. F 5 2. 'round'

and were delighted with the result (see Fig. 11.1(b)). They then experimented with different size 'sticks' for their lollipop, discovering that 'F1' gave a small stick, while 'F7' gave a larger one. They then reversed the order of the commands:

1. 'round' 2. F 5

and discovered this led to an 'upside down' lollipop (Fig. 11.1(c)).

Susannah and Corrie carried on experimenting, and suggested to each other that they did a 'silly shape'. Corrie suggested:

1. 'square' 2. 'triangle'

which produced the famous 'bugged house' illustrated in Papert's book *Mindstorms* (Fig. 11.1(d)). When this appeared on the screen both girls spontaneously burst into applause!

Apart from noting their obvious interest and enthusiasm, we did not carry out a systematic evaluation of this initial work with the Nursery children. However, Muriel Slade, the teacher in charge of the Nursery, spontaneously remarked that the Turtle work had had a marked effect on the confidence of these children: she described them as being 'full of beans' after their sessions. This effect was not simply the result of being taken out of the Nursery to play games with an adult (to which children in our Nursery were accustomed), but seemed specific to the Turtle.

This brief experience suggested that even children of preschool age might benefit from using the Turtle and, in particular, might be encouraged by it to carry out mathematical exploration. Further confirmation for this view came from another local Nursery teacher, who visited the Department and watched our children using the Turtle. Some time later we were able to lend her a spare Turtle and Concept Keyboard for a few weeks. She carried out some interesting and intensive work with a group of three boys from her Nursery class. At the end of this period she concluded: 'I firmly believe that it is a worthwhile project and that we should try to see a way of using it within the classroom.'

12

The Craigmillar Logo Project

THE SCHOOL AND THE CHILDREN

Craigmillar is a severely deprived council estate on the southern outskirts of Edinburgh. Its social problems are profound. According to a recent report, the rate of eviction is the highest in the city, the rates of illegitimate birth and attempted suicide are three times the city average, and the number of children in care is four times the city average. Intervention by the Royal Scottish Society for the Prevention of Cruelty to Children is seven times the city average. Craigmillar has the highest proportion of children receiving free school meals, and the incidence of mental handicap is three times as high as in other areas.

This deprivation is reflected in the levels of educational achievement of children living in Craigmillar. The reading age of children in Craigmillar Primary School is on average fifteen months behind their chronological age, and for many children the difference is three years or more. In a study carried out in 1980, we found that even the three-year-olds in the nursery class were already a year behind children from more affluent areas in their understanding of basic number concepts (Hughes, 1981). This level of attainment exists despite the constant efforts of an enthusiastic and dedicated teaching staff in the school itself.

Why, then, did we choose to set up our project in Craigmillar Primary School? Our main reason was simply a feeling that if Logo proved to be as beneficial to young children as we hoped, then the Craigmillar children deserved to be amongst the first to enjoy the benefits. Unlike children in more affluent areas of Edinburgh, very few children in Craigmillar have access to computers at home, and so are in danger of being placed at a further disadvantage compared to other children. We also wanted to reverse the more usual practice of carrying out innovatory projects with highly motivated and articulate children. The history of educational innovation contains many examples of projects which have worked with more able children, but failed in areas like Craigmillar. Our motto was — 'If it works in Craigmillar, it'll work anywhere'.

Our work with the children lasted from September 1983 to March 1984. It was preceded and followed by periods of three to four weeks when we carried out various tests on the children and undertook in-depth interviews with the school staff. During the summer term 1984, the class teacher continued the work in the classroom using the Screen Turtle and the Concept Keyboard.

The initial sample of children consisted of all the Primary 2 children in a Primary 2/3 composite class. At the start of the school year this totalled 17 children (11 boys, 6 girls) with a mean age of 6 years 1 month (range: 5 years 7 months – 6 years 5 months). The sample was soon reduced to 15 children, as one boy left the school shortly after the project began, and another boy was absent through illness for most of the first term. A third boy missed about a third of the project through illness, but he was included in the various evaluation measures. The sample was augmented by a girl who joined the class about half way through the project. The remaining 14 children were reasonably regular attenders throughout the period of the project.

The children's initial knowledge concerning computers was virtually non-existent. We had expected thenm to have some ideas, perhaps derived from 'Star Wars' or 'Dr Who', and we were surprised by how little the chidren knew, and by their apparent lack of interest in the subject. In the course of preliminary interviews the children often confused computers with radios, televisions and videos. One girl, when asked if she would like to draw a picture of a computer, made an elaborate sketch of the interviewer's tape recorder.

A few children demonstrated some idea of the appearance of a computer:

Interviewer: What is a computer?
John: It has buttons down the bottom, and a big round thing on top. ... You press the buttons and it does the letter ... on the thing. ...
Interviewer: What does it do?
John: You press the buttons and it asks you questions.
Interviewer: Have you seen one?
John: On telly.

One of the most interesting conversations resulted from a digression about robots, in an otherwise unfruitful interview. Steven had drawn a picture which he said was of a robot:

Interviewer: What is a robot? [Pointing to the picture which Steven had drawn] It looks like a man.
Steven: It's a kind of monster.
Interviewer: Is it alive?
Steven: Yes. It can't go dead.
Interviewer: Why?
Steven: It's made of metal.
Interviewer: How does it come alive?
Steven: 'Cos it's got stuff in its tummy.

All the children were briefly assessed on their knowledge of left and right, their knowledge of basic shapes such as squares and triangles, and their basic number skills. Seven out of sixteen children could identify their own left and right hands, but only one could correctly work out which was the left and right hand of someone sitting opposite. Virtually all the children could recognize and name basic shapes, although few could give adequate instructions as to how to draw them. Their competence with numbers up to 10 was adequate, but beyond 10 was extremely vague; in general their number competence was well below average for their age.

The main method of working was to withdraw the children from the classroom in groups of 2 or 3 to a room which was given over to the project. Occasionally the children were withdrawn individually. At least one member of the research team was in the school for 3 or 4 mornings a week, and the children were withdrawn for sessions of 15–25 minutes. Over the 5 months of the project the children received on average 24 sessions each, totalling from about 6 to about 10 hours of contact with the computer.

The children all proceeded through the same sequence of software options, from the simple STARTER option to PEN, then SHAPES and lastly SCREEN. These options will now be described in more detail.

STARTER

STARTER is a very simple set-up for introducing the child to the Turtle. The screen is not used, and the Turtle is controlled by means of the five buttons 'F', 'B', 'L', 'R', and 'H' which are used one at a time and are immediately effective. They correspond to five commands as follows:

- 'F' moves the Turtle forward one Turtle length (about 28 cm);
- 'B' moves it back one Turtle length;
- 'L' turns it 90 degrees to its left;
- 'R' turns it 90 degrees to its right;
- 'H' causes it to 'hoot' (play the first few bars of "Scotland the Brave").

The buttons are arranged on the Concept Keyboard as shown in Fig. 12.1.

The children's initial reaction to STARTER was one of considerable enthusiasm: 'It's barrie' (a local term of extreme approval) was frequently

Fig. 12.1 — Keyboard layout for STARTER.

heard at this time. In the early days they were easily distracted, and occasionally they got over-excited. Our initial approach was to stand back and let them explore the system, but it quickly became evident that some constraints were needed on the children's behaviour.

The children rapidly grew proficient at operating the keyboard. The distinction between 'F' and 'B' was learnt almost at once, but the distinction between 'L' and 'R' presented more widespread problems: indeed, some children were still confusing 'left' and 'right' at the end of the project. 'Hoot' was extremely popular at the start, but later became less so and eventually was used only in particular situations, such as when the Turtle had been successfully steered to the desired location.

The main activities for these sessions consisted of driving the Turtle around the floor. The children built houses and 'garages' for the Turtle out of cardboard boxes, leaving large doors for the Turtle to move through. We also joined up these houses and garages by roads drawn on the floor with chalk: the children steered the Turtle along the roads trying to keep within the chalk lines.

After four sessions with STARTER, the children were shown how to use the Turtle's pen. The keyboard was expanded to include two new keys, 'U' (pen-up) and 'D' (pen-down). At the same time the effects of the 'L' and 'R' commands were changed so that the Turtle turned 30 degrees rather than 90 degrees, to give the children finer control.

The Turtle was placed on a large plastic-coated board, allowing the children to draw either directly on to the board or on to a sheet of paper placed on the board. Initially, activities were suggested in which the Turtle moved from one location to another leaving a 'trail' behind it: attention was not focused explicitly on drawing pictures. In fact the children themselves soon started to attend to the shapes being produced by the pen.

A typical session at this stage is one involving Neil and Kevin. This was in fact their first introduction to the pen. Neil drew, without planning, a shape consisting of two sides of a triangle (see Fig. 12.2(a)). Kevin pointed out that

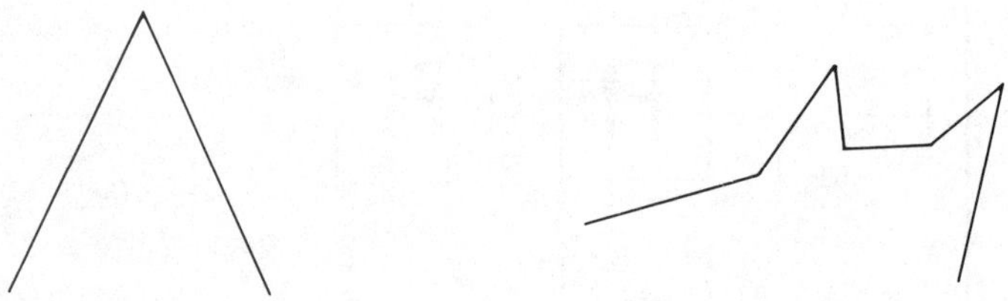

(a) (b)

Fig. 12.2 — Drawings produced using STARTER. (a) Neil's tent. (b) Kevin's 'cat's ears'

it was 'like a tent', and told Neil to 'make a triangle'. Neil joined up the base of the triangle, and Kevin suggested he continued it along to make 'a kite' (i.e. with a string); Neil did not do this, however. When it was Kevin's turn he drew the pattern shown in Fig. 12.2(b), pointing out that it was like 'a cat's ears'. On Neil's second go he produced the 'square shape' shown in Fig. 12.3(a). Kevin took over, saying that it was a 'house' and tried to draw a 'roof'. He decided at the end it was more like a 'coal bunker' see Fig. 12.3(b).

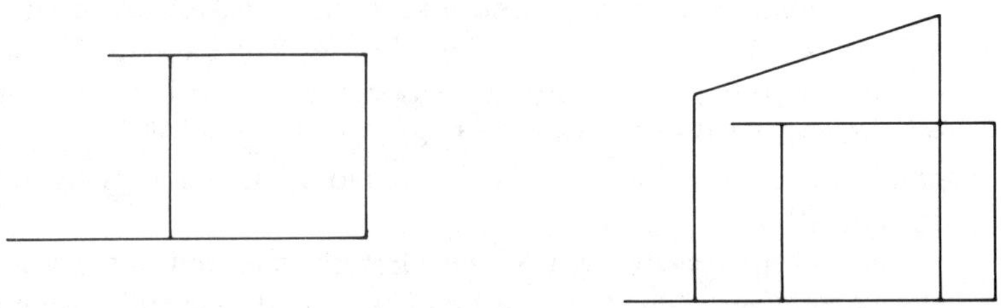

(a) (b)

Fig. 12.3 — More drawings using STARTER. (a) 'Square shape'. (b) 'Coal bunker'.

Clearly the children's drawings at this stage were very rudimentary: however, what comes through clearly from these remarks was their readiness to see familiar objects in the outline drawings produced by the Turtle.

PEN

After approximately 6 sessioons with STARTER, the children progressed to the second stage, the option called PEN. This essentially introduced them to programming in its simplest form: the Turtle commands were no longer immediately effective, but were first displayed on the screen. This showed a picture of a Turtle with a large 'thought-bubble' above its head: if the child pressed 'F', the letter 'F' appeared in the bubble (see Fig. 12.4). In addition,

Fig. 12.4 — Screen display for PEN.

the PEN option introduced scaling for the first time. That is, the child could now combine movement commands with distance commands, as in 'F2' or 'B3', or rotate it through an angle such as 'R1' or 'L4'. There could be only one such command in the thought-bubble at a time, and it could be changed simply by entering a different one. When the child was satisfied with the contents of the bubble, the command was transmitted to the Turtle by pressing the new 'GO' key (see Fig. 12.5).

The children's immediate reaction to the extended Concept Keyboard was enthusiastic. One child even remarked 'It's like a real computer now!'. However, it took them some time to master the new scaling (this combination of enthusiasm and difficulty seemed to occur always when the system was expanded). At this time they limited themselves to very simple shapes.

After a few sessions with PEN we introduced the idea of planning. We asked the children to make an initial plan of the drawing they wanted to produce with the Turtle, and followed this by discussing with them how the plan might be put into practice. As usual the children reacted enthusiastically to this new idea, but their initial plans bore little resemblance to anything the Turtle might have been able to produce. However, their ability to produce realistic plans gradually improved.

The children undoubtedly made a great deal of progress during a week in early November, when Cathie Potts, a member of the research team,

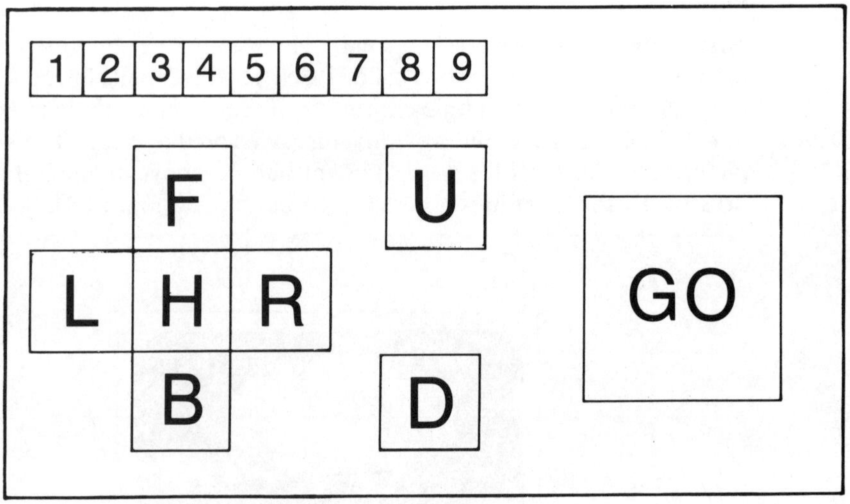

Fig. 12.5 — Keyboard layout for PEN.

worked with them individually, helping them to produce plans and execute them. As Cathie herself noted at the time:

> Once you have on paper a drawing which you want to reproduce with the Turtle, you are in a good position. From then on you share a common goal with the child. . . . Previously I had often felt that to be helpful towards completing a picture could be made difficult by the fact that I might be imposing my idea of what, for example, a

'Turtle car' might look like.

An example of this kind of planning comes from a session with Nadine. She decided she would like the Turtle to produce a drawing of a cat, and drew the plan shown in Fig. 12.6(a). This drawing was quite complicated but, when she was asked which parts the Turtle could be used for, Nadine pointed to the simple outline, recognizing the Turtle's limitations. She followed her plan as far as the cat's back and ears but ran out of time and so finished it off by hand. The final figure is shown in Fig. 12.6(b).

Another example of planning comes from Julie. She had seen some pictures of owls on her way along the corridor and wanted to try one with the Turtle. The first plan she produced (see Fig. 12.7(a)) did not show consideration of the limitations of the Turtle. After a brief discussion about the kinds of lines and shapes she had made with the Turtle before, Julie produced her second plan (See Fig. 12.7(b)). This time she pointed out that the bird was made up of 'a square, a triangle and straight lines'. When she eventually came to operate the Turtle, Julie followed her second plan quite carefully and was delighted with the result. She added the bird's face and wing by

(a)

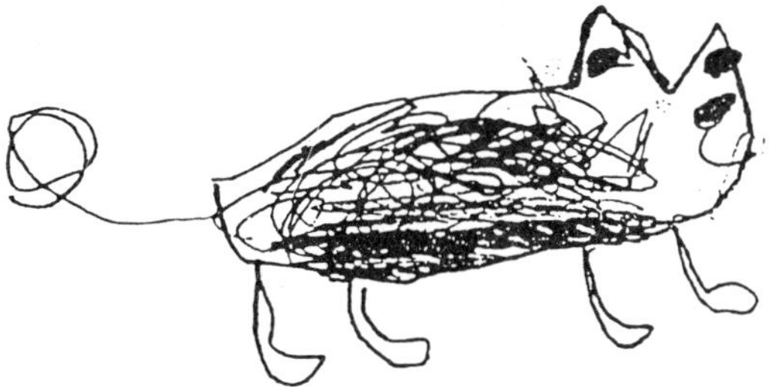

(b)

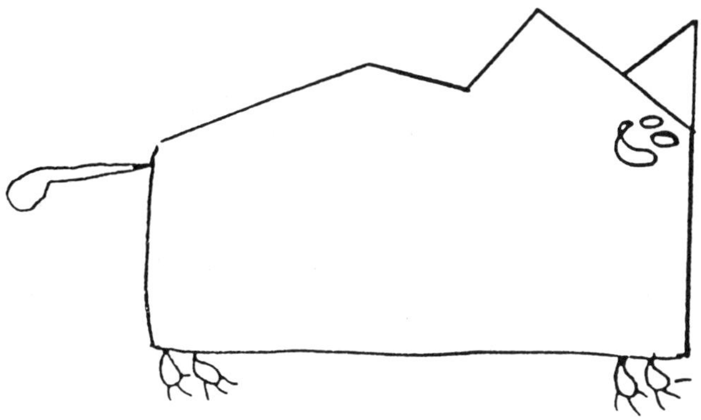

Fig. 12.6 — Nadine's cats.

hand, and when she was asked whether or not owls had ears, she added the rather large ears. The final figure is shown in Fig. 12.7(c).

SHAPES

At the end of November we felt the children were ready for the next stage. We briefly tried introducing some of the children to the concept of 'procedure building' (entering sequences of commands), but they showed little interest in this. Instead, we expanded the system to allow them to use four pre-programmed shapes, triangle, square, star and what was usually referred to as a 'round' (initially our fourth shape was a hexagon, but we replaced this with a 12-sided figure). At first the prepared shapes were of standard size, but we then moved on a stage further, enabling the children to scale the shapes (see Fig. 12.8 for SHAPES keyboard). Thus 'square 2' produces a square with side 2 units, etc.

The facility of using prepared shapes was obviously attractive to the children and it allowed them to produce more complex drawings. At the same time, it introduced new difficulties. For example, the children had to learn the direction in which the Turtle would lead off when starting any

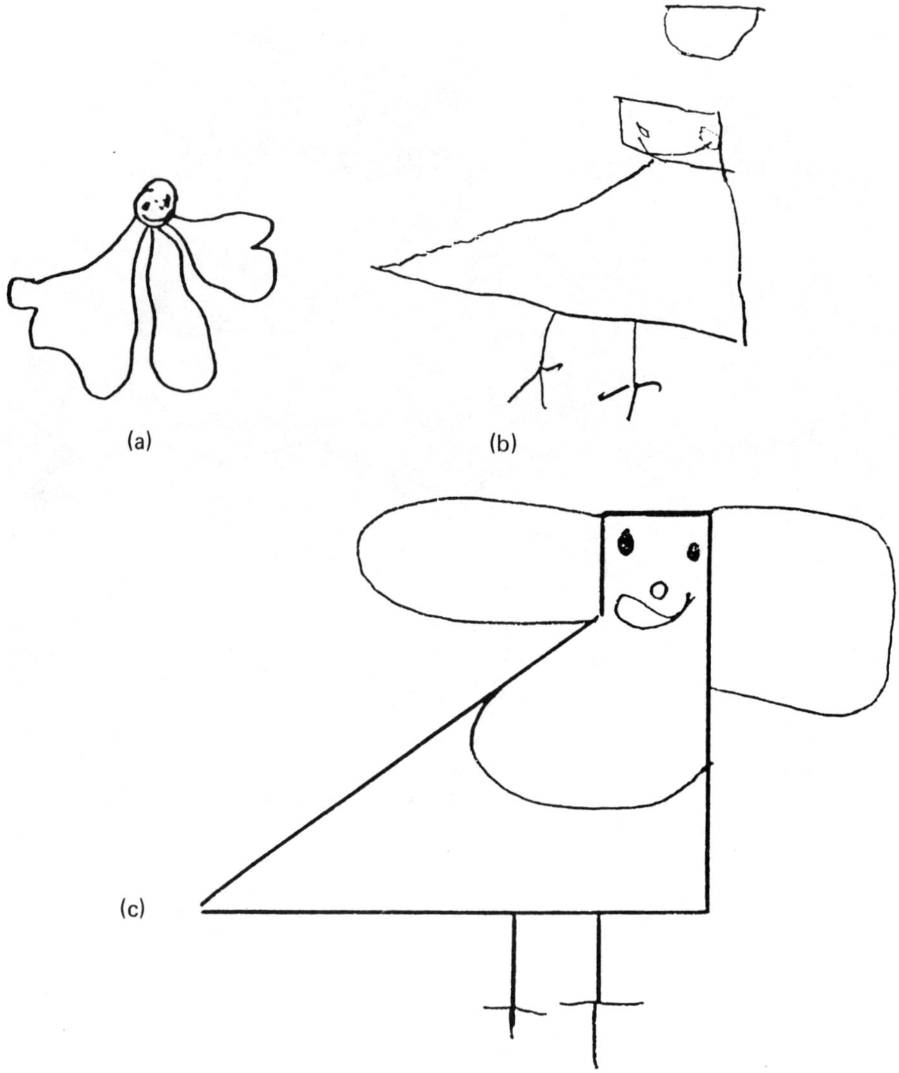

Fig. 12.7 — Julie's owls.

particular shape, and the subsequent orientation of that shape, and had to plan accordingly.

We shall now give a detailed summary of one particular session using SHAPES. The summary gives a good illustration of the kinds of mathematical discussion that the Turtle could elicit. The session, which took place just before Christmas, involved three boys — Kevin, John and Andrew — working with Cathie Potts.

The idea of drawing a snowman comes from John. Kevin is also keen, but Andrew less so; eventually, however, he agrees. Somewhat unusually they

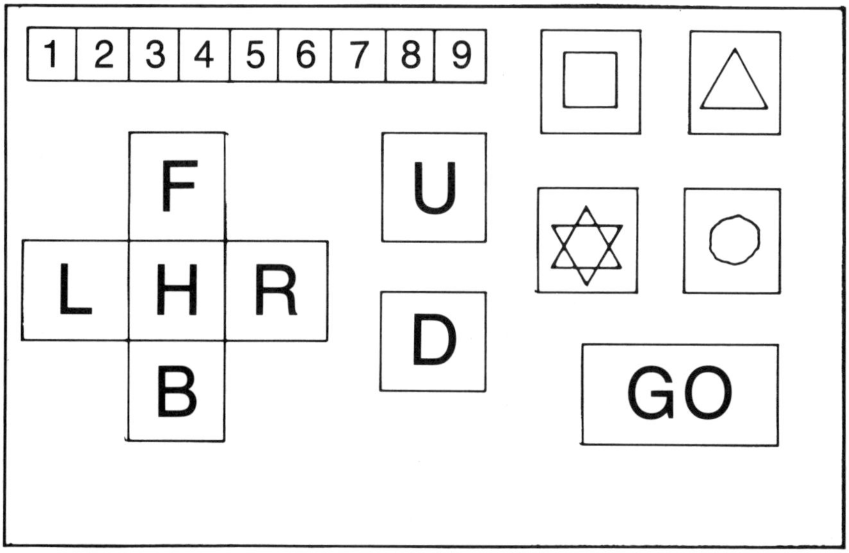

Fig. 12.8 — Keyboard layout for SHAPES.

do not make a plan — it is clear that they all share the same idea of what a snowman should look like. See Fig. 12.9 for the result of their session.

Kevin starts at the keyboard, with Andrew sitting next to him and John standing behind him. Kevin puts the pen down, and Cathie puts the Turtle in place on a large sheet of paper. Kevin enters 'round' and '5'. The Turtle sets off, and Andrew says 'You're making a hexagon.' John asks if it's the tummy or the head, and Kevin comments that 'That makes a big one.' John says 'That can be the body. You could easily make that be the head if you want, but it'll go off the paper.'

When the circle is finished, both Andrew and John point out that the pen will have to be raised before the Turtle can be moved to the top of the snowman's body to make the head. Kevin raises the pen. He now needs to turn the Turtle 90 degrees to the right, but puts in L3 instead. Andrew corrects him, and together they manoeuvre the Turtle into position to start the snowman's head.

The first attempt to draw the head is 'too wee', but luckily the pen is still up. Andrew repositions the Turtle for the head, and they discuss what size to make it:

John: How big do you want it to be? One or two? ... Dinnae press 'nine'.
Andrew: Three.
Kevin: Then 'GO'

The Turtle draws the head.

John: That's quite a good snowman.

Kevin: Buttons.
Andrew: It'll be easy to make buttons, you just press 'one' and 'circle'.
Kevin: Hexagon, hexagon.

Andrew takes over from Kevin at the keyboard. He manoeuvres the Turtle to the middle of the snowman's body, although he has to be reminded that he must put the pen up first.

Kevin: Make buttons now.
Cathie: What size are you going to make them?
Kevin: Small.
Andrew: Mmmmmm ... one.
John: I say about one.
Kevin: It's very good!

The first button is finished. Andrew says 'We'll need to make its hat', and John suggests 'I could do a pipe'. With Cathie's help, Andrew manoeuvres the Turtle into position for the second button and sets off the Turtle. However, he has forgotten to lower the pen, and has to repeat the command. The second button is also a success.

John now takes over at the keyboard. He takes the Turtle up to the top of the snowman's head to draw a hat, but exact positioning of the Turtle using commands is impossible as it needs to be in the middle of a Turtle length. Cathie positions it by hand, and the boys discuss what kind of hat to draw: Kevin suggests a triangle, Andrew a 'circle hat' and John a square. Cathie points out that a square hat could be given a brim. The Turtle is not facing in the right direction to start a square, and Andrew tells him to 'Turn it three'. John in fact keys in 'R2', and Andrew tells him, 'You need another one.' John adds this and puts the pen down.

Cathie: What size of square do you think?
Kevin: Small.
John: Two.
Andrew: A small one — three.
John: [looking at Cathie] What do you say?
Andrew: Shh, listen to what Cathie says. What number?
Cathie: I don't know. John, I think you should decide, it's your hat.
Kevin: Should take four.
Cathie: I think that would be too big, wouldn't it?
Kevin: You've never took four before.
Andrew: Do three. ... It'll be too big.
John: I say two.
Cathie: If you say two, put in two, John. Put in two. [The square is drawn] That's fine, isn't it?

John takes the Turtle backwards to produce part of the brim. The boys discuss how to produce the rest of the brim

Andrew: [points the way the Turtle should go] Try two.
John: [points the same way] Right.
Andrew: Try two. [The Turtle goes forward, but not far enough. Andrew kneels beside it.] How many do you need for the brim? [Asks Cathie] One?
Cathie: About the same again, I think. About another one or two.
Andrew: Two. ... One. [He holds up one finger to John.]
John: Try one.
Kevin: And then do another one?
Cathie: Well, it's got to go to the corner, and one past. ...
John: Right, two. ... Will I go forward? [Turtle goes forward two.]

They discuss with Cathie what else they can do. They decide that eyes would be too small, and John says he will draw them by hand. They agree to draw a triangle for a nose. The Turtle is placed (by hand) in the middle of the snowman's face, but Andrew has worked out it is facing in the wrong direction: 'We'll need to put it down that way.'

They discuss what size the nose should be. John wants a big one, and says 'Three', but Andrew suggests 'Two' and then 'One'. The Turtle draws a triangle of size one, but John is disappointed: he wanted a bigger nose.

Andrew wants to finish the snowman by drawing a pipe. Cathie asks John how he thinks Andrew will do it.

John: Along. ... A square on the end ... easy.
Andrew: [draws a long stem for the pipe] That's an awfy big one. I done three.
Kevin: I don't believe it!
Cathie: Just a wee square.
Andrew: What nunber? One?

Cathie nods, and the Turtle draws the square bowl for the pipe. The bell goes, but John stays behind to draw in eyes, mouth and smoke.

SCREEN

For the final six sessions of the project the children transferred from the floor Turtle to the Screen Turtle. Of these six sessions, two were an introduction to SCREEN, two involved the use of a printer in conjunction with it, and two were part of our final evaluation. However, the children continued to use SCREEN in the classroom after the end of the project.

The keyboard layout for SCREEN was the same as that for SHAPES (see Fig. 12.8) except that the 'H' ('hoot') button was replaced with a 'C' ('clear') button which, when pressed, would erase whatever was on the screen at the time.

As usual, the introduction of a new facility was greeted with enthusiasm but caused difficulty. Some of the children were dismayed at the Floor Turtle not being used, but most were interested in what the Screen Turtle could do.

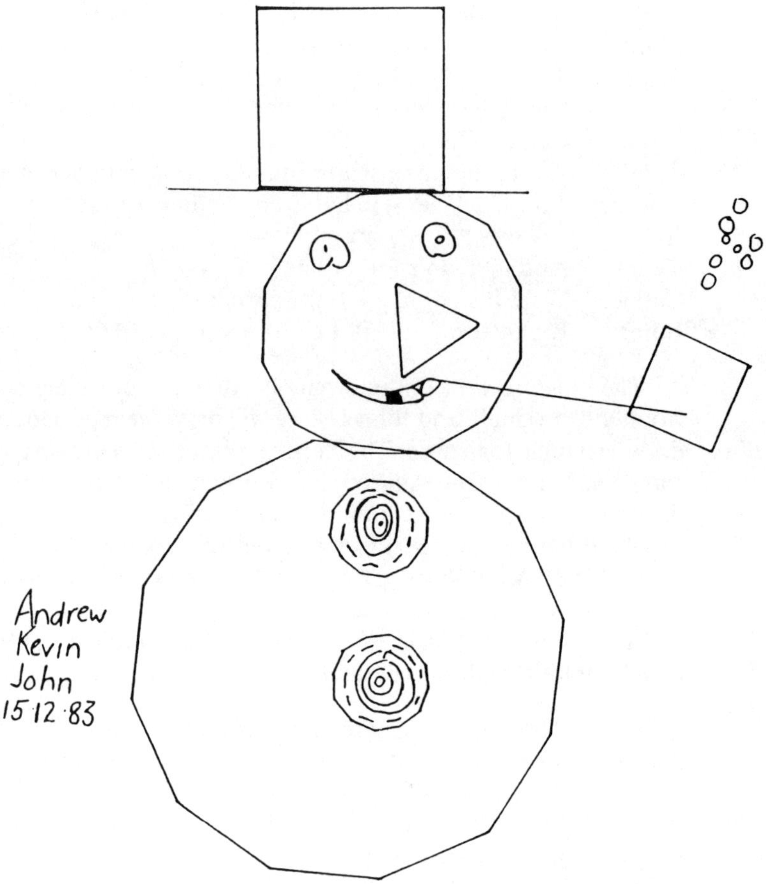

Fig. 12.9 — Snowman.

There was a new interest in experimenting with patterns — for example, exploring the effects of repeatedly entering the same shape but with different parameters. Fig. 12.10, for example, was produced by the se-

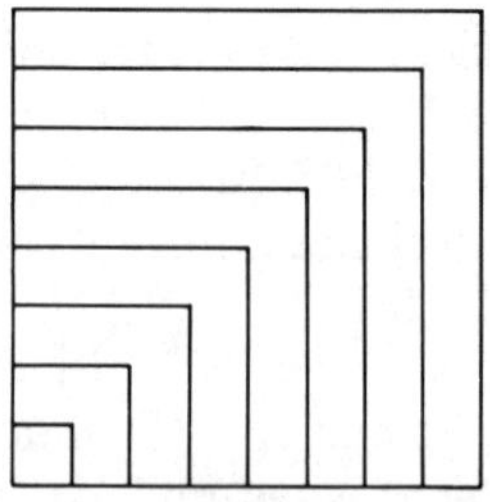

Fig. 12.10 — Squares within squares.

quence 'square 1', 'square 2', 'square 3', etc.

As expected, it took the children some time to grasp the meaning of the commands 'U' and 'D' ('pen-up' and 'pen-down') in the SCREEN context.

However, there was little sign of the expected confusion between 'U/D' and 'F/B' (up/down the screen). Nor was there any problem in referring to the enlarged cursor on the screen as a 'Turtle'.

The introduction of the printer was again an exciting development, allowing the children to print-out copies of their screen drawings, and they quickly learnt to operate it themselves.

With SCREEN the whole process of producing drawings was much faster, and they could experiment more in their drawings. However, the use of detailed plans fell away in this period. Instead, the children started to write out successful programs in a book in the classroom, and the idea of a drawn plan was replaced by that of a set of instructions.

An example of a session with no drawn plan, but showing the flexibility of working with SCREEN, is one with Lynn and Kelly-Ann. Kelly-Ann started first, saying that she was going to draw 'a rose . . . a big circle with wee ones all round'. She got as far as one large circle with two on its edge when she commented that it looked like a teddy bear (see Fig. 12.11(a)) and changed her plan. Kelly-Ann and Lynn worked together on the positioning of the cursor to start the next large circle for the head. They both understood the value of trying a shape with the pen up at first, to check that it would appear in the right position. With one ear left to add, and several minutes into playtime, Kelly-Ann asked, 'Could it no be a very old teddy and one ear fallen off?' Lynn was more persistent and added the final ear (see Fig. 12.11(b)). The two girls took their print-outs to the classroom to colour in.

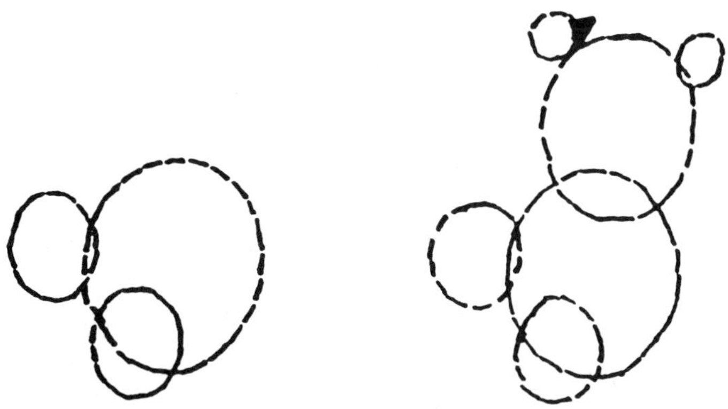

(a) (b)

Fig. 12.11 — Teddy.

Jason and Nadine wanted to draw a street. Jason started by attempting to draw a house on the screen. At first he made a triangle inside a square, producing the famous 'bugged house' figure mentioned by Seymour Papert (see Fig. 12.12(a)). He then realized the need for a 'forward' command between the shapes, and made the triangle on top of the square, but still with the wrong orientation (see Fig. 12.12(b)). When Jason had worked out how to position the triangle as a roof, he had yet to realize that, for it to fit, its parameter had to be the same as that of the square. He worked it out by making successively bigger triangles.

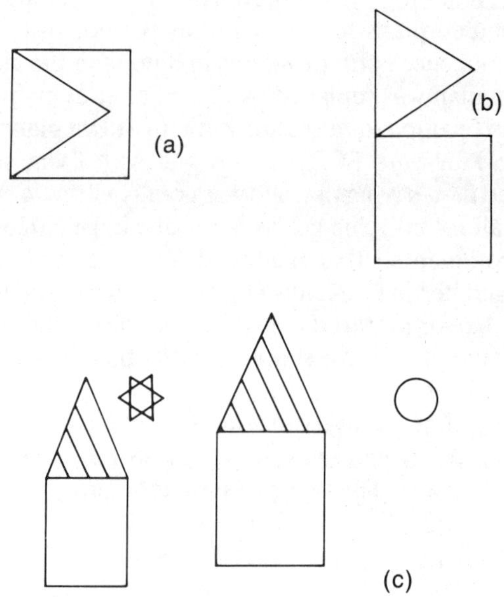

Fig. 12.12 — Houses.

Nadine made her house correctly at her first attempt, having learned from Jason's mistakes. She added stripes to her roof 'so they match', and then added the moon. Jason completed the picture with a star (see Fig. 12.12(c)).

13

Evaluation of the Craigmillar Logo Project

While there are many enthusiastic claims made about the value of Logo, there are few clear-cut research findings to support these claims. A review by Ross and Howe (1981) concludes that the results of research in the previous decade have been 'more encouraging than discouraging, but only mildly so'. Interestingly enough, this review suggests that the positive effects of Logo have been made mostly in the area of mathematics: thee is virtually no evidence that Logo helps children in the more general area of problem-solving. Pea and Kurland (1984), in a similar review, criticize both the methodology and the assumptions of previous studies, and conclude that the available evidence fails to provide clear-cut answers. Finlayson (1984), in attempting to reconcile the lukewarm research evidence with the enthusiasm of teachers using Logo, stresses the need for educational programs and evaluations to be much more specific in the issues they are addressing. Such requirements seem reasonable, given the relative novelty of the whole area.

Our own view is that it is still very much an open question as to which aspects of children's development — if any — are likely to be significantly affected by Logo, and our aim in this evaluation was to build up a broad picture, from as many sources as possible, of the effects of the Logo experience. In this chapter we will report on the evidence we collected from the following sources:

(a) informal observations of the children using Logo;
(b) interviews with members of the Craigmillar staff about the effects of the project;
(c) systematic pre- and post-testing of the children using the British Ability Scales — a standardized abilities test;
(d) systematic assessment of the children's Logo competence at the end of the project;

(e) individual interviews with the children about the way the computer and the Turtle operated.

(a) OUR OWN OBSERVATIONS

At the start of the project we were very much plunging into the unknown. the method we were developing could have worked well with the children, or they could have been a total failure. As it happened, they worked even better than we had imagined.

This does not, however, mean that every session with the children flowed along smoothly and effortlessly. The reality was very different. Progress was often frustratingly slow. There were times when the children appeared to have made very little progress from one week to the next, or even to have gone backwards, forgetting something that we thought they had grasped much earlier. These periods were, however, usually ended by a sudden leap forward, as the children moved unexpectedly on to a higher level of performance. It was only when we looked back to the beginning of the project that we realized just how far the children had progressed in a relatively short space of time.

We were particularly struck by the following aspects of the children's behaviour.

(1) Motivation and concentration

We expected that the children's interest and enthusiasm for the Turtle would be high at the start, but might rapidly die away. In fact, they maintained their interest in the Turtle throughout the period of the project. Moreover, their ability to work in a persistent and focused manner at the keyboard increased considerably as the project wore on. At the start, they were excited by the novelty of the equipment, and the initial sessions were usually chaotic. At the end, the children were much calmer, and capable of sustaining their work at the keyboard for periods of 30–45 minutes.

(2) Ability to co-operate

We were impressed by the children's increasing ability to work together on Turtle drawings: to agree on a joint project, and to discuss and argue about what they might do. Naturally, some children worked better together than others, but whereas it is often claimed that computers are isolating and tend to cut children off from each other, the evidence of our project suggests that in fact computers have a tremendous potential to encourage interaction and discussion between children.

(3) Use of mathematical language

One of the original intentions behind the project was to introduce mathematical ideas and terminology in a context that was very real for the children and, as the examples given earlier demonstrate, the children did indeed make considerable use of mathematical language. Getting the Turtle to carry out a particular operation frequently led to discussion of shape, size and distance. Particularly striking was the way the children became able to estimate whether to send the Turtle a small distance or a longer distance, and there was frequent discussion about whether a line had to be a 'four' or a 'five' for example. There was also discussion as to whether turns were 'left' or 'right', although these concepts were still presenting difficulties to some children at the end of the project.

(b) INTERVIEWS WITH THE CRAIGMILLAR STAFF

The second source of evidence for the effects of our project came from in-depth interviews with members of the school staff. Their comments supported many of our own observations. The gains which they felt had accrued to the children fell into the following areas:

(1) Concentration and absorption

There was general agreement that the Turtle work produced high levels of concentration from the children, and that this was almost entirely self-motivated:

> It's great, they really love it... The things that they're doing with the Turtle are much more advanced than when they started off. You know, they're really thinking. Like Craig, when he did that ship, he really thought about what he wanted and he explained it all — 'That's the ship, and that's the rail, the safety bars to hold you on, so you don't fall off'. And just the thought that's gone into the pictures...
>
> (Class teacher)

> Their concentration — that's the thing that bowled me over the first time I saw it all. It still does. The number of them who stick at it and stick at it. It's amazing. And it's quite complicated — 'How do we get the Turtle up there?', 'Where do you want it to go?' — you can almost *feel* the amount of concentration going in.
>
> (Headteacher)

(2) Increased mathematical understanding

There was also agreement that the Turtle work helped children's mathematical understanding, particularly in the area of number and shape. The Class teacher had noticed:

> ... increased awareness of properties of shape — for example, that squares all have sides the same length. They now understand much more about angles, they know the meaning of big and small angles. They're also much better at estimating... and the layout of their work's much neater.

(3) Language and discussion

The staff were also struck by the way the Turtle work had stimulated the children's use of mathematical terminology in particular, but also their general communicativeness:

> The children talk a lot about the Turtle work, especially when they're just back from a session... They talk to each other more about the Turtle than they do about anything else.
>
> (Class teacher)

> I certainly think their language has improved... If they hadn't been planning what they were doing then they wouldn't have been having the kind of communcation they were having with each other... It's the sharing and discussion between themselves on the best way to go about something, which you don't even get in much older children in this school.
>
> (Headteacher)

(4) Confidence and maturity

There wee also observations about effects on the children's general demeanour:

> I think they're more secure in themselves, more confident if you like...
>
> (Headteacher)

> They're somehow more reasonable than the children of the same age who haven't had this experience...
>
> (Assistant Head Teacher, Lower school)

> They're much better now at working together in the classroom...
>
> (Class teacher)

The interviews witht he staff also revealed that the project had overall made a very favourable impression. Two comments in particular were very gratifying to the research team. The first came from the Assistant Head Teacher of the Upper school, who had previously been teaching BASIC to this Primary 7 pupils:

> I've seen the value of Logo, it's widened my vision of it... Last year I was teaching BASIC and one of the things that I have changed by mind about, during the course of the project, is the value of BASIC as a thing to teach primary pupils. In future I will not begin to teach children BASIC, I will move right on to Logo.

The second came from the Headteacher, who admitted that the project had made her think again about the capabilities of some of the Craigmillar children:

> It's taught me something, it really has. Because one labels these children... That's given me something to think about...

The interviews also revealed some of the deeper issues associated with the arrival of computers in primary schools. For example, there was a great deal of dissatisfaction with the kind of software currently available, especially concerning the limited amount of understanding which any piece of software might be promoting:

> There's a danger of people seeing the computer as doing things that I can't see it doing... There's so little progression in much of what's available... I'm concerned about the amount of knowledge or skills the children will get out of it.
>
> (Assistant Head Teacher, Lower school)

In addition, there was concern about the establishment of a new hierarchy in schools, with those who understood computers at the top and other teachers feeling left out:

> It's the in-thing to be good with a computer at the moment...
>
> (Assistant Head Teacher, Lower school)

> What frightens me is the way people who know about computers have a power over other members of staff — how there's a whole race of computerate teachers emerging and they're a kind of high priesthood...
>
> (Assistant Head Teacher, Upper school)

Undoubtedly, though, most of the teachers would agree with the Headteacher, who readily admitted that her own attitude to computers had really changed:

> I now see tremendous uses for it, both for the more able children and the pooer ones... It's not just a question of keeping them quiet, it's a question of stimulating them in a way that a teacher just can't do.

(c) ASSESSMENT OF CHILDREN USING THE BRITISH ABILITY SCALES

At the beginning and end of the project the children were all individually assessed on the British Ability Scales (BAS). This is a standardized instrument which consists of a number of independent tasks (sub-scales). Each sub-scale measures a specific ability. In addition, the scores on the sub-scales can be compiled to produce an overall IQ score.

We used the following sub-scales:

Matrices. This is a non-verbal reasoning test similar to the Raven's Matrices. The children are shown a series of patterns with one part missing. Their task is to fill in the missing sections.

Similarities. The tester reads out the names of three objects (e.g. 'orange, strawberry, banana'). The child has to mention another object that goes with them (e.g. 'apple'), and then supply a class name that explains why they all go together (e.g. 'because they are all fruit').

Block Design. The child is given a set of patterned cubes, and asked to put them together to produce various designs. Each design has a time limit related to the degree of difficulty.

Copying. The child is asked to copy a series of designs on to plain paper.

Digit Recall. The child is asked to repeat a series of digits, spoken at half-second intervals (e.g. '2, 3, 7, 4, 6).

Basic Number. This scale contains a range of items which test understanding of concepts such as more/same/less, relative size and quantity, simple addition and subtraction , and counting in tens and units.

Naming Vocabulary. The child is shown pictures of objects and asked to supply their names. Objects include: fish, scissors, cup and saucer, spade, ring, watch, triangle, van, sink/washbasin, jar, chain, robin, switch, scales, funnel, compass.

All the sub-scales have demonstration items to ensure that the child understands the task. All are discontinued when the child fails a certain number of items consecutively.

The test was administered to the children by a postgraduate student in the Psychology Department who had trained as an educational psychologist. She was not involved in the project and knew nothing about the individual children.

The child's score on each sub-scale is related to their age and then compared with the scores achieved by a large (nationwide) sample of children of the same age.

Because the BAS relates the child's performance to their age, it is particularly useful for assessing gains the children might make over a period of time. Obviously, we would expect children to perform better on the sub-scales when re-tested several months later, simply because of the passage of time. However, we were looking for evidence that the children had improved over and above what we would expect with time. Thus, if the project had produced no effects on the children, their scores would remain the same. Any significant differences between the initial and final scores can be assumed to be due to the project.

Table 13.1 shows the children's initial and final performance on the BAS. The last column shows whether the differences are statistically significant. 'NS' means the difference is not significant. However, if the probability (p) of a chance result is 0.05 or less, then the difference is statistically significant.

As Table 13.1 shows, statistically significant gains were found on the following sub-scales: Block Design, Digit Recall and Basic Number. These scales were all specifically concerned with either number or shape. In other words, we can conclude from these figures that the Logo project had a significant effect on the children's understanding of number and shape, but not on other aspects of their development.

A separate analysis was carried out for the boys and the girls (see Tables 13.2 and 13.3). This analysis revealed significant gains for the boys on the Basic Number and Block Design sub-scales. It also found, for the first time, an overall significant gain in IQ. However, there were no significant gains for the girls on any of these measures, including IQ.

These findings require some qualification. In particular, it should be noted that the sample size is very small, especially for the girls ($n = 6$). In the case of the girls, it would be extremely hard to show any significant gains in such a sample, and there are suggestions of gains on the Digit Recall and Basic Number sub-scales. There is, however, the suggestion of a decrease in the girls' Matrices scores; this is puzzling, and is a major contributor to the lack of an overall gain in IQ.

All the same, despite these qualifications, there is a strong suggestion from the BAS scores that the girls in our project did not do as well as the boys, although it must be borne in mind that they started from a higher initial level. Nevertheless, the finding of differential effects is worrying, and indicates at the very least the need for a further study with a much larger sample of both boys and girls.

(d) ASSESSMENT OF CHILDREN'S LOGO COMPETENCE

At the end of the project, each child was individually shown three pictures produced by the Screen turtle — a flag, a balloon with a star on it, and a face (see Fig. 13.1(a)) — and asked to produce on the screen pictures exactly like them. The children readily complied with this request. Their drawings were subsequently scored (out of 5) for their resemblance to the original, taking into account the degree of adult assistance they had received. Thus a score of 5 showed that the child had produced and correctly placed all the main features, with no help from the adult; 4 showed a similar achievement with some help; 3 showed that the child had, unaided, produced but incorrectly positioned the main features; 2 showed the same level of achievement with some adult assistance; 1 showed some attempt to produce the main features; 0 meant no attempt had been made. Agreement of the scoring was reached through discussion; no reliability data is available.

The children's performance on these tasks varied considerably. At one extreme were children such as Andrew (Fig. 13.1(b)), who were able to produce figures close to the originals with a minimum of adult intervention.

In the middle of the range were children such as Nadine (Fig. 13.1(c)), who were able to draw a flag with little trouble and make some progress in producing the elements of the balloon and face. The poorest performer was Nicola (Fig. 13.1(d)), who could produce the flag only with considerable guidance. Even with help from the adult Nicola had difficulty with the balloon. She did not attempt the face.

Table 13.1 — Pre- and post-testing using British Abilities Scales, all children ($n = 15$)

	Pre-	Post-	Diff.	Significance
Sub-scale				
Matrices	51.3	53.6	+2.3	NS
Similarities	40.5	47.9	+7.5	NS
Block Design	45.9	54.9	+9.0	$p<0.05$
Copying	32.8	37.9	+5.1	NS
Digit Recall	48.3	60.6	+12.3	$p<0.05$
Basic Number	28.1	50.5	+22.4	$p<0.005$
Vocabulary	47.8	55.1	+7.3	NS
Computed IQ	98.1	103.0	+4.9	NS

Table 13.2 — Pre- and post-testing using British Abilities Scales, boys only ($n = 9$)

	Pre-	Post-	Diff.	Significance
Sub-scale				
Matrices	45.0	57.9	+12.9	NS
Similarities	31.9	45.9	+14.0	NS
Block Design	54.9	67.9	+13.0	$p<0.02$
Copying	33.2	46.2	+13.0	NS
Digit Recall	39.5	49.5	+10.0	NS
Basic Number	27.9	56.2	+28.3	$p<0.005$
Vocabulary	53.5	64.4	+10.9	NS
Computed IQ	94.2	103.0	+8.8	$p<0.05$

Although the boys performed better than the girls on this task (mean scores out of 15: boys 9.6, girls 7.0), the difference was not statistically significant (Mann–Whitney U-test). It seems unlikely, then, that the sex

Table 13.3 — Pre- and post-testing using British Abilities Scales, girls only ($n = 6$)

	Pre-	Post-	Diff.	Significance
Sub-scale				
Matrices	60.8	47.2	− 13.7	NS
Similarities	53.3	51.0	− 2.3	NS
Block Design	32.3	35.3	+ 3.0	NS
Copying	32.2	25.5	− 6.7	NS
Digit Recall	61.5	77.2	+ 15.7	NS
Basic Number	28.3	41.8	+ 13.5	NS
Vocabulary	39.2	41.0	+ 1.8	NS
Computed IQ	103.8	103.0	− 0.8	NS

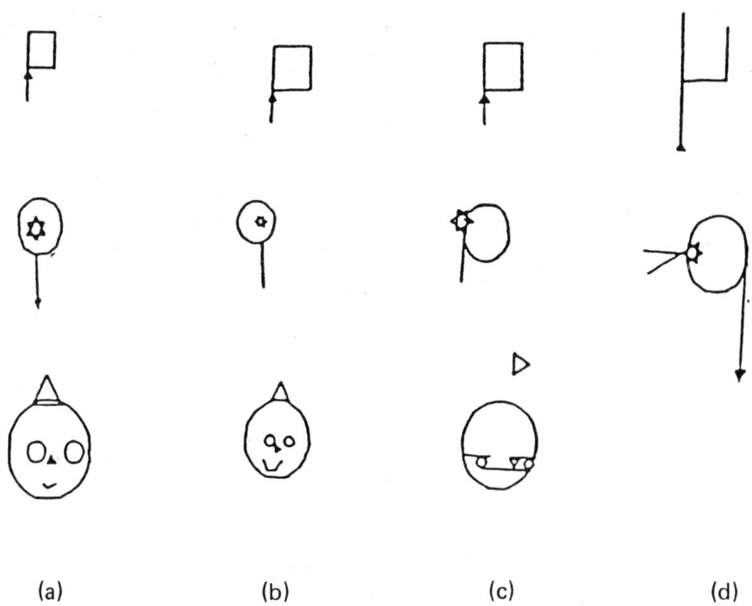

Fig. 13.1 — Test of Logo competence. (a) Standard. (b) Andrew. (c) Nadine. (d) Nicola.

differences found in the BAS scores were caused by sex differences in the children's Logo competence. Indeed, there was very little correlation between whether the children did well on Logo and whether they gained significantly on the BAS.

(e) THE CHILDREN'S UNDERSTANDING OF THE COMPUTER AND TURTLE

At the end of the project the children were interviewed individually about their understanding of how the computer and the Turtle operated. They were asked to draw the equipment (which was not in the same room), and a picture such as the Turtle might have made.

Fig. 13.2 shows two typical drawings of the equipment. Anthony's

(a) (b)

Fig. 13.2 — Children's drawings of equipment. (a) Anthony. (b) John.

drawing (Fig. 13.2(a)) shows the Turtle and the Concept Keyboard. On the keyboard he has represented 'F', 'L' and 'R' ('B' was probably used less often by the children while drawing). The control lead of the Turtle is shown ending in space, with no connection to the keyboard (the lead was held suspended by a spring from the ceiling to keep it out of the way of drawings). John's drawing (Fig. 13.2(b)), however, shows a clear understanding that the Turtle and the rest of the equipment are connected, and he has drawn the computer's video screen above the Concept Keyboard. Again, John's representation of the keys on the keyboard is very limited and probably reflects the relative importance to him of some of the functions.

When the children produced samples of 'Turtle drawings' (either spontaneously, or after encouragement from the interviewer) they were asked to explain the commands which would be necessary to direct the Turtle to produce the drawing. There were considerable differences in the children's abilities to perform this task. Alex, for example, was able to break down his drawing of a person into its component parts (see Fig. 13.3(a)), but was not

Fig. 13.3 — 'Programs'. (a) Alex. (b) Andrew. (c) Neil.

able to represent these components with Turtle commands. The other two examples show greater competence. The program to draw the man was written by Andrew (Fig. 13.3(b)). It contains instructions for the head, body, face, arms and hands. No account is taken for the need to move the Turtle with the pen up to position these features correctly, although the order in which they should be drawn is completely clear — apart from the fact that almost everything is written backwards! With the exception of the numbers associated with the turns, Neil's program to draw the car is very accurate indeed (Fig. 13.3(c)). The Turtle starts drawing from the bottom right corner of the picture, and it is quite possible to follow the train of thought which produced the program. The program concludes with two 'rounds' used as wheels.

In addition to this exercise based on the children's own drawings, the interviewer drew freehand a standard picture of a house as shown in Figure 13.4(a). This was made up of a square with a perfectly fitting triangle on top.

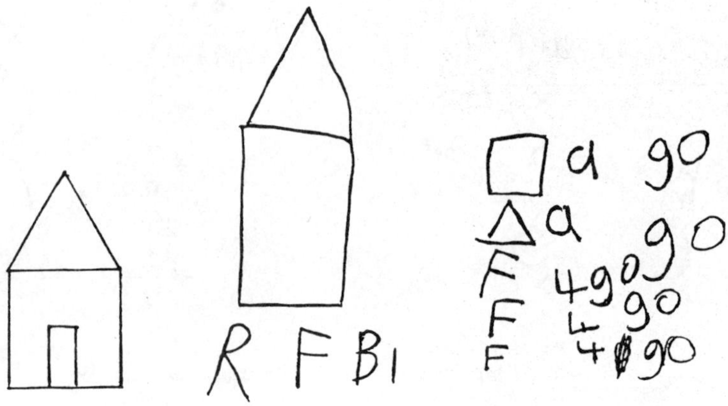

(a) (b) (c)

Fig. 13.4 — 'Houses'. (a) Standard. (b) Jason. (c) Nadine.

The centrally positioned door was half the height of the wall and obviously rectangular in shape. Each child was asked to explain how they would tell the Turtle to draw it.

As before, there were considerable differences in the success with which the children were able to deal with this task. Some children responded merely by redrawing or copying the given drawing, giving a verbal description of what was going on: Jason's attempt (Fig. 13.4(b)) is little more than this. In contrast, Nadine's program (Fig. 13.4(c)) shows a clear understanding of how she would first use the prepared shapes 'square' and 'triangle' to produce the outline of the house, and then draw the three sides of the door using 'forward' commands. Unfortunately, she omits the turns!

All the programs produced by the children in the course of this inverview were later scored, in relation to either the standard house or their own drawings. The scores were based on the presence, in their attempts, of a number of different features, such as: a representation of the drawing as a number of sequential steps; an understanding of the role of the numeric parameter within the command; a recognition of the need for 'invisible' steps within such a program, i.e. an understanding that there was more to producing the drawing than listing the features actually produced by the pen; and the general accuracy of the program in terms of distance, angle and the ordering of the commands.

The children's scores on this interview correlated significantly with their scores on the standard Logo test. That is, children who did well with an actual computer were also good at discussing how they would produce a particular hypothetical figure. However, there was no correlation with their gains on the British Ability Scales. Nor was there any significant difference between the boys and girls on this measure.

14

Where next?

The project described here lasted for only eight months. Nevertheless, within that short space of time it produced a number of clear findings.

First, it demonstrated that the Logo approach to using a microcomputer is indeed feasible with children as young as six years. Moreover, our preliminary work with pre-school children suggests that the approach can be validly used with four-year-old children, and possibly even younger. It seems reasonable to conclude that the version of Logo which we have developed is a useful addition to the early education of young children.

The project, however, demonstrated more than just the feasibility of this approach; it also demonstrated that many of the claims made about Logo for older children are just as true for younger children. Even for children as young as six, Logo provides a doorway into a world that interests them and stimulates them. Our own informal observations, as well as those of the teaching staff of Craigmillar Primary School, reveal that when working with Logo the children are capable of sustaining an interest and involvement far beyond anything we have seen elsewhere. The Turtle makes them think, it makes then discuss, and it makes then explore: it makes them set themselves problems and discuss how to solve them, and it makes them use mathematical language in a way that is real and meaningful for them. The fact that one activity has all these effects on the children is indeed impressive, and suggests that we need to give far more attention to this approach as a way of helping young children to think and to learn.

The standardized assessment of the children also revealed significant gains in areas where we had expected the approach to have the most impact. Their scores on the BAS sub-scales concerned with number and shape showed significant gains, and the lack of significant gains on other scales suggests that the effect is not simply a 'blanket' improvement between pre- and post-tests. It is somewhat disturbing, however, that these gains seem to be limited to the boys in the sample, although, as we pointed out earlier, the number of girls taking part in the study was very small, and the boys were starting from a lower initial level than the girls. No doubt further work will reveal whether there is a genuine difference between boys and girls in the way they take to this particular approach, or whether our findings are simply the product of our particular sample.

The project has also raised a number of questions which need to be explored further.

First, we still need to discover how easily the Logo approach can be used in the normal infant school classroom. We were working under ideal conditions, extracting children in small groups to work with the Turtle in a room given over for that purpose. Moreover, the children were helped by the constant attention of a member of the research team working with them. If an infant teacher were to use our set-up in her classroom, she would inevitably encounter difficulties, such as how to organize access to the computer. Does she allow the children free access, which may lead to the computer only being used by those children who are particularly interested and excited by this approach? Or does she organize her classroom work so that the computer is being used by all the children in some sort of rotation?

There are other problems too, connected with the use of the Floor Turtle in a classroom. It requires a certain amount of space, and the children working with it must neither be distracted from what they are trying to achieve nor interrupted needlessly in carrying it through to its conclusion. The teacher will not be able to give the children the kind of attention that was possible from us: with the rest of the class to consider, the time available for the Turtle, and for intervening when help is needed with it, may be very limited. Nevertheless, we believe that, although transfer to normal classrooms will cause problems, these problems are in principle open to solution.

A second problem which needs to be addressed concerns progression within the Logo language. The children in our sample were taken through an ordered sequence of activities which increased in complexity as their understanding of the system grew. If this approach is to be maintained then this sequence must be continued. The children in our study were, by the end of the project, ready for more complex versions of Logo, but at that time there was no version available for the BBC Micro that would easily have mapped on to the work they were already doing. This problem has been partially resolved by the arrival of full versions of Logo for the BBC, but the problem of dovetailing the two parts of the language still remains.

The third problem opened up by our work concerns the relationship of the Logo work to other kinds of mathematical work taking place in the classroom. In many ways the Logo approach is based on principles very similar to those underlying modern approaches to early mathematics. For example, Logo encourages children to explore: it puts them very much in control of their learning, and it frees them from the routine of carrying out set arithmetical exercises. Yet there are still discrepancies between the Logo approach and most maths schemes available at present. The most obvious of these is in the area of shape, where the Logo approach introduces a number of quite complex shape notions concerned with angle and distance much earlier than they are introduced in the children's maths schemes. Our experience suggests that this early introduction of ideas about shape is something which the children can cope with. The implication, therefore, is that we need to consider how to adapt existing maths schemes to the Logo approach, rather than vice versa. At a deeper level, there may be discrepancies due to the emphasis which Logo places on number as a measure of distance, rather than as a measure of quantity. Again, our work suggests that linking numbers to concrete distances helps children understand more about number, but it may help them to understand only some aspects of number. Much more thought needs to be given to these problems.

The fourth areas where further work is required is in the development of appropriate teaching strategies. Within the Logo movement this is an area of considerable debate. There are some who believe that Logo is, in principle, a system that can be used by children without any help from the teacher at all. At the other end of the spectrum, there are those who believe that it must be used in a structured way, as part of a set of exercises designed by the teacher. The approach we adopted lies midway between these two. We encouraged the children to set their own goals and to draw plans of what they wanted to achieve; we then gave them whatever help was necessary to achieve these goals. We need, however, to give much more thought to the precise kinds of intervention needed when a teacher is using Logo. Are there, for example, common and recurring conceptual problems with which children require adult help? It is only through using Logo in many more classrooms, and studying the effects on the children, that we will be able to develop an understanding of these issues.

Finally, we need to give more thought to the question of how to evaluate the Logo experience. In our work so far we have used a number of approaches, including informal observation of the children's behaviour and their performance on standardized tests. Both these approaches suggest encouraging gains. However, we feel that there is room for a much more detailed and careful analysis of exactly what children are learning when they are working with the Turtle and how this learning is taking place. Again, this analysis can only come as a result of the frequent use of Logo with children of different ages, abilities and backgrounds.

REFERENCES FOR PART III

Berdonneau, C. (1985) 'An approach to mathematics in a LOGO-like environment with kindergarteners.' Paper presented to the LOGO and Mathematics Education Conference, London.

Finlayson, H. (1984) 'Mathematical strategies and concepts through turtle geometry.' Paper presented to BPS Developmental Conference: 'AI, IT and Child Development'.

Hoyles, C., Sutherland, R. and Evans, J. (1985) 'A preliminary investigation of the pupil-centred approach to the learning of LOGO in the secondary school mathematics classroom.' University of London, unpublished report.

Hughes, M. (1981) 'Can preschool children add and subtract?' *Educational Psychology,* **1**, 207–219.

Hughes, M. (1986) *Children and Number.* Oxford: Basil Blackwell.

Lawler, R. W. (1985) *Computer Experience and Cognitive Development: A child's learning in a computer culture.* Chichester, Ellis Horwood.

Metz, M. (1985) 'LOGO, little children and large numbers.' Paper presented to the LOGO and Mathematics Education Conference, London.

Noss, R. (1983) 'Starting LOGO: Interim Report on the Chiltern MEP LOGO project.' Unpublished report, Microelectronics Education Programme.

Papert, S. (1980) *Mindstorms: Children, computers and powerful ideas.* Brighton: Harvester.

Pea, R. D. and Kurland, D. M. (1984) 'On the cognitive effects of learning computer programming.' *New Ideas in Psychology* **2**, 2, 137–168.

Perlman, R. (1974) 'TORTIS — Toddlers own recursive turtle interpreter system.' *LOGO Memo 9,* MIT, Cambridge, Mass..

Ross, P. and Howe, J. A. M. (1981) 'Teaching mathematics through programming: ten years on.' In Lewis, R. and Tagg, D. (eds.) *Computers in Education.* Amsterdam: North-Holland.

Vaidya, S. R. (1984) 'Making LOGO accessible to preschool children.' *Educational Technology,* July, 30–31.

Index